Enough

Fermentation, Time, and the Conditions of Transformation

Brian Golbere

Luludelulu Press

Contents

Introduction

I make sauerkraut, kimchi, and kombucha. I've brewed beer, baked sourdough bread during lockdown, and I enjoy natto, which says something about my tolerance for processes most people find unpleasant. I know what a ferment smells like on day three versus day ten. I recognize the moment the kombucha mother begins to form, how the brine clouds as lactobacillus takes hold, and the particular sourness that indicates the culture is working versus the different sourness that signals something went wrong. These are not metaphors to me. They are Tuesday.

This framework isn't a literary device. I've never seen the world the way most people around me do. The patterns I observe, how time feels, the sense that coordination is more fundamental than the objects being coordinated — these aren't realizations I developed over time. They are how the world has always shown itself to me.

So I started searching. Years of searching. The growth phase. A mix of curiosity and frustration as I tried to understand how I comprehend. That search eventually led to a formal theory. The theory generated implications. The implications kept branching, connecting, and deepening. And at some point, the growth phase began to feel like a jar being filled forever but never sealed.

I had had enough. Not everything, not the entire account, but enough to get started. Enough substrate, enough structure, and enough of the pattern visible so that remaining in the evaluation phase—refining, extending, and preparing to communicate eventually—was no longer growth; it was delay. The theory didn't need to be perfect; it needed a vessel. Similarly, the sauerkraut doesn't require the cabbage to be flawless. It just needs the jar of cabbage with salt, sealed.

This book is that vessel. It is a crossing. Not because writing it was courageous or because the ideas are groundbreaking, but because at some point I stopped judging whether the world was ready for what I see and began the fermentation: here is the shape, as clearly as I can make it, sealed in a form that either resonates with what your fitness function already recognizes or doesn't. I can't control which. I can only do what practice requires — commit to the vessel, maintain the temperature, show up — and let the process unfold.

The book is my answer to my own question. Not a theory. A fermentation of a theory. The complexity here wasn't just in the ingredients. It's what emerged when the loop ran through my particular substrate long enough to produce something I couldn't have predicted from the parts. I've been inside fermentations my whole life — in the kitchen, in thinking, and in the slow process of trying to find a form for something I could taste but not yet name. This book is what that process produced. Not what I planned. What it became.

If that resonates with something you already sense but haven't found the words for — if the shape of things described here matches something you've been quietly noticing — then the book is effective. Not because I taught you something, but because your internal system recognized a pattern it was already running. And if that recognition is genuine, it belongs to you, not me. I just sealed the jar.

Prologue: The Jar

There is a jar on your counter. Its shape is older than you think — shaped over generations by hands that learned the process through doing, with its form carrying what language can't express. It's been sealed for days. You can't see what's happening inside. Nothing, it seems. The cabbage looks the same. The brine is cloudy. You fight the urge to open it.

You will not open it. Not because you're patient — patience has nothing to do with it. Because opening the jar is the one thing that stops the process. The transformation happening in that jar requires the thing you have been trained, by every system you participate in, not to provide: unmonitored time.

The world you operate in has an opinion about sealed jars. It would like you to open it. Measure the progress. Optimize the timeline. Share the update. The world you operate in runs on the opposite of fermentation — it runs on extraction, which requires access. Access to your attention, your coordination, your time. The jar is an affront to extraction because extraction needs the lid off. It needs to see what's inside because that's the resource.

Everything in this book is about what happens when you keep the lid on.

Not forever. Not as withdrawal. As the specific, deliberate act of providing the conditions under which transformation — yours, not the version someone else is monitoring — becomes possible.

The presented world will tell you two things about your jar. The first: you should be making something better. You should be authoring the outcome, manifesting the result, sculpting the fermentation into what you intended. The second, for when the first fails: the jar is empty, the process isn't working, you need a different substrate, a different recipe, a different life. These are the same voice. One sells you agency you don't need. The other sells you rescue from the despair that the first one manufactured.

Both require the lid to be off.

Both are wrong.

What's actually happening in the jar is not something you author or something that fails. It is a coordination loop — a living process running through your substrate, reorganizing it from within. The culture doesn't consume the substrate and leave. It runs its own cycle, and the material gets reshaped around that cycle. The visible products — the tang, the fizz, the preservation — are outputs. The real transformation is structural: complexity that wasn't in the original ingredients, produced not by addition but by a living loop running through them long enough to change what they are.

You cannot get this through extraction. You cannot press the grape harder to get wine. The transformation requires time and a self-sustaining process running through the substrate — older than language, older than intention, operating by dynamics that don't require your supervision and can't survive it. What comes out is not what went in, and it is more than what went in, and it cannot go back.

But only if you leave the jar sealed.

Part One: Substrate

What you're working with

Chapter 1

Enough

Enough is where we always start.

Everything that transforms requires time, temperature, and the willingness to seal the jar.

You already know this. You've known it longer than you've had language for it — the felt sense of a meal that has become satisfying rather than just continuing, a conversation that has said what it needed to say, a day that didn't need to be more than it was.

That recognition is not passive. It's an act of navigation. You have been reading a threshold your whole life, in both directions — *is there enough here to continue? And has this reached enough, for now?* — and acting accordingly. The assessment runs constantly, mostly below awareness, calibrating against something you couldn't articulate but have always been using.

That something is a fitness function. An ongoing evaluation of whether what you're doing is sustaining what needs to be sustained, transforming what's ready to transform, and releasing what has completed.

You didn't learn it. It came with you.

What you may have learned, over time and through particular environments, is to distrust it.

The distrust isn't accidental. It's cultivated. Every system that extracts from you has an interest in you not trusting your own fitness function — because a person who knows when they have enough is a person who stops consuming. Stops scrolling. Stops optimizing. Stops buying the next framework, the next solution, the next version of themselves that will finally be ready.

The presented world runs on two myths, and both require your fitness function to be offline.

The first myth says you are the sculptor. You create your reality, you manifest your outcomes, you author your life. If that isn't producing what you want, you aren't trying hard enough, visualizing clearly enough, or manifesting with sufficient intention. This myth sells you infinite responsibility for a process that was never yours to control — and then sells you tools to manage the anxiety that infinite responsibility produces.

The second myth is the sculptor's shadow. If you can't sculpt your reality, then you must be the clay. Powerless. Determined. A victim of forces too large to resist, along for a ride you didn't choose. This myth sells you safety. Protection. Someone else's structure, because yours couldn't hold.

Both myths suggest the same idea: don't trust the signal within you. The one that says *this is enough to work with*, or that it has reached enough for now. Because trusting that signal makes both myths unnecessary, and each has its own business model.

The deeper error underlying both is that reality is static. Something to be sculpted or surrendered to. A given. The sculptor and the clay are both operating on a stage that was already built, fighting over who controls it.

Reality is not a stage. It's a fermentation. Dynamic, reproducing, constituting itself through processes you are embedded in, whether you know it or not. Your fitness function is not an opinion about a fixed world. It is your participation in an ongoing one that you evaluate as viable that feeds into what persists.

That's a different kind of agency than sculpting. Harder. More consequential. Less dramatic. And it doesn't need the lid off. Its existence is proof.

The fitness function evaluates. It reads the threshold, assesses what's present, and compares what is to what could sustain. That reading is continuous and mostly invisible — the background hum of a system monitoring its own viability. Evaluation is not crossing. You can evaluate forever. Many people do.

A crossing is the moment you act on the evaluation. The irreversible moment. The decision that cannot be changed back. Not the deliberation — the step. Not the assessment of whether the water is deep enough — the jump. After a crossing, the world is different from what it was before, however slightly, and you cannot undo it by re-evaluating.

This matters because the world presented to you is very good at keeping you in evaluation mode. Gather more information. Consider more options. Assess risk. The growth phase is endless. Evaluation feels like agency — you're thinking, weighing, preparing. But it constitutes nothing. No time is produced. No transformation begins. The jar is being inspected, not sealed.

Crossings are how time gets constituted. Every genuine decision, every irreversible commitment, every moment where you stop assessing and start acting — these produce the quanta of time that the clock merely counts. A life rich in evaluation but poor in crossing is a life thin in constituted time, no matter how busy it appears.

Reading this book is, for now, an assessment. That's okay. It's part of the growth phase, and it's essential.

Here is the thing about the grape.

It already contains everything required to become wine. The yeast is on the skin. The sugars are there. The water, the acids, the precursors of every flavor compound that will emerge — all present before fermentation begins. Nothing is added. The loop just runs.

This reframes everything quietly but completely. The extraction impulse — pressing harder, adding more, optimizing the process — comes from the assumption that what you have isn't sufficient. The grape needs help becoming wine. But the grape is already enough. Not enough in the consolation sense — *make peace with less than you wanted*. Enough as a precise claim about the nature of the substrate: it already contains what the transformation requires.

The feeling that something is missing is exactly what leads to the interventions that hinder fermentation. Overly aggressive temperature control. Premature racking. Adding sulfites that kill the native yeast before they can do their job. Each intervention stems from a lack of trust in the substrate, and each results in a wine that falls short of what the grape already had the potential to become.

The role of time isn't additive either. Time doesn't add anything to the grape; it is the medium through which the loop runs. You can't speed it up without changing what you get, not because speed is bad, but because the coordination has its own natural pace. Rushing fermentation doesn't make wine any faster; it just produces something else.

Which means "enough" and "time" aren't separate concepts here. Enough is what the substrate already is. Time is what allows that sufficiency to manifest. Neither works without the other, but neither is the missing ingredient — because nothing is missing.

That's probably the deepest thing this book has to say.

Enough runs in two directions.

In one sense, this acknowledgment is that what you have is enough. Not perfect, not complete in the sense of needing nothing more, but enough — the foundation contains what the next transformation requires. Recognizing this is not passive acceptance. It is the act that makes sealing the jar possible. You cannot commit to a fermentation you believe is missing a critical ingredient. The seal requires trust in what's already present.

In the other direction, enough is completion. Something reaches enough—a project, a relationship, a phase of life, or a version of yourself — and completes. Not fails. Not gets abandoned. It completes. It has done what it can do at this temperature, in this vessel, with this substrate. Recognizing that completion is an act of agency — reading the threshold in the direction most of us were never taught to look.

The two directions are the same reading. Sufficiency at the start and completion at the end are both part of the fitness function—evaluating what exists, what can be achieved, and what the substrate can sustain. *For now* matters. Enough is not permanent. It is linked to the current stage of the process. What completes at one stage becomes the foundation for the next. The completion is real without being final. The jar is sealed, the transformation occurs, and what endures is the culture, refined and more complex than what was initially present.

Enough is a physical condition, not a quantity. Not an idea.

Every structure that exists — from a particle to a person to a civilization — persists because it surpasses a minimum coordination threshold. Below this threshold, nothing remains unified long enough to develop into anything. Above it, things endure. There is also an upper limit: the point at which complexity surpasses what the structure can sustain, where adding more causes coherence to break down rather than grow.

The entire domain of transformation exists between these two limits. Fermentation happens there, too. Below the lower limit: nothing happens, the substrate remains inert. Above the upper limit: the process becomes too hot, destroys the complexity it was building, produces something volatile, or collapses. Between them lies the interval in which existing resources can express themselves. Not where you accumulate toward sufficiency. Where what has always been present gets the right conditions to transform.

Temperature determines your position within that range. In any living system, temperature reflects the rate of coordination — how swiftly decisions are made, how rapidly crossings are settled, and how much time is being constituted per unit of activity.

Which means time is not the container in which fermentation happens.

Time is what fermentation produces.

More on that later. For now, the substrate is already sufficient, the interval is real, and you have been navigating both your whole life with more precision than you've been given credit for.

The quality of that navigation is not about finding the right point and holding it.

A healthy heart doesn't beat like a metronome; it varies—constantly adjusting its rhythm based on conditions. The fluctuation between beats signals health. When the heart loses this flexibility, it narrows and becomes rigid, like a metronome. High heart rate variability indicates the system can respond effectively. Low variability suggests it's locked in place.

Enough works the same way. It's not a number you find and defend. It's the responsiveness of your reading — the ability to feel "enough to begin" one day and "enough for now" the next, to notice when the threshold shifts because the process has shifted, to adjust between the lower and upper limits with awareness of what is actually happening instead of following a fixed plan.

Someone locked on "not enough" — always below the threshold, always scrambling, never sufficient — has low variability. Their fitness function remains at a single reading regardless of conditions. Someone locked on "too much" — overwhelmed, past the limit, adding complexity faster than they can integrate — is stuck in the opposite direction. Both are rigid and metronomic.

High variability means your fitness function has room to adapt. It can accurately measure the current threshold, rather than relying on the previous one or a prescribed value. This variability isn't something you accomplish through effort; it's a downstream indicator—showing how much of your coordination ability is genuinely available instead of being used up by unfinished tasks, extractive couplings, and the background noise of a system operating at someone else's temperature.

The threshold itself isn't fixed; it shifts as the process evolves. What was sufficient substrate yesterday might not be enough today because fermentation changes what's needed. What was excessive input last month could be just right now, as the system has developed the complexity to handle it. The threshold is a property of the process, not something imposed from outside. This means navigating it effectively requires the same thing fermentation does: ongoing contact with actual conditions. Not a map. Not a rule. Just the live, responsive reading itself, connected to what's here.

Your fitness function already accomplishes this. Every crossing in your assembly history shaped it—making it unique to you, not generic, calibrated by the specific stresses, consolidations, and cultures that built the system you're running. No one else's fitness function interprets the threshold the way yours does. That uniqueness is not a problem to fix; it is the instrument.

The book simply asks you to notice it.

Chapter 2

Assembly History

You can only ferment what you actually have.

Not what you planned to build. Not what you intended to accumulate. Not the substrate you would have if different crossings had been made, different environments encountered, different cultures received. What you actually have is the product of your assembly history — and that history is not the length of your life.

It is the length of every successful crossing that got you here.

Millions of years of them. Every threshold that was crossed and sustained. Every coordination that lasted long enough to become structure. The cell that developed a membrane. The cells that understood each other. The nervous system that integrated fast and slow processing into something capable of navigation. The social structures that encoded successful behaviors into culture so each generation didn't have to start from scratch. The languages that condensed understanding into a transmissible form. Each of these was a crossing — irreversible, consequential, real — and every one that succeeded became a foundation for the next.

You are the current surface of that accumulation. Not the endpoint. The current surface. Every relationship that left something behind, every understanding that consolidated or failed to, every phase of growth that built real complexity or extracted without building — these are the most recent layers. But they sit on top of the ones that built the capacity for relationships, for understanding, for growth in the first place.

Assembly history is not the background. It is the ingredient list. And the ingredient list goes deeper than you can narrate.

This is the constraint that often goes unacknowledged in the literature of self-improvement, which tends to act as if the substrate is infinitely malleable — as if you can choose to ferment differently, as if the quality of what emerges depends solely on intention and technique. It doesn't. It depends on what you're working with. And what you're working with was not created by you. Most of it was assembled long before anything resembling you existed. Your contribution is the most recent layer — real, consequential, yours — resting on millions of years of substrate that was already load-bearing when you arrived.

This is not a limitation to overcome. It is a useful thing to know.

———————————

You did not choose your starter culture.

The family you were born into, the language that structured your early thinking, the nervous system calibrated by environments you were too young to evaluate, the historical moment that set the tone of your formation — these are not conditions you operated within. They are the organisms doing your fermentation. They are the culture you're running.

Everything you've subsequently built is based on that culture. The specific organisms. The particular microbial signature of your history. No two fermentations are the same because no two starter cultures are identical, even when the ingredients appear similar.

This is entrainment. You are coupled to processes larger than yourself — biological, familial, cultural, historical — and those processes are setting your temperature, your rate, the chemistry of your transformation. The coupling is not optional. It came with the substrate.

What is optional — just barely, and only with sufficient accumulated complexity — is noticing what you're entrained to.

Noticing itself is a phase transition, needing threshold conditions. It cannot be forced. But when it occurs, something becomes accessible that wasn't before: not the ability to escape the culture you're in, but the ability to recognize it as a culture. This means, at minimum, that you can ask whether it's producing the transformation you're experiencing.

Assembly history has two components that are easy to confuse.

The first is what was consolidated: the relationships that built true coordination capacity, the understanding that grew into genuine depth, the crossings that were fully paid for and created a foundation rather than debt. This is the load-bearing material. It doesn't seem like much from the inside — what is integrated becomes invisible, which is what makes it structural.

The second aspect is what wasn't completed. Processes that were interrupted before phase transition, understandings that calcified before they could evolve, relationships that remained active past their natural endpoint. This material still exists — still consumes coordination capacity to sustain, still occupies the active process queue — but it's not adding complexity. It's a burden without a gain.

Both are substrates. Both are what you're working with.

The first question of any honest inventory is: what has actually solidified? Not what should have, not what you remember working on — what is truly load-bearing now, integrated to the point of invisibility? And what remains in the active queue, still being maintained, still waiting for the acknowledgment that it has reached enough and can be released into transformation?

You cannot will the second category into the first. But you can stop interrupting the process that moves things from one to the other.

Assembly history is not a closed ledger. It is still assembling.

Right now, as you read this, crossings are happening. Some are completing, while others are consolidating into structure you'll barely be able to distinguish from yourself in six months. Some are failing to reach the threshold and will stay in the active queue until they are recognized as complete or released.

This is easy to overlook because the word *history* suggests something finished. But the assembly is a process, not a record. The millions of years of successful crossings that created your substrate — those weren't finished either at that time. They were the current surface of their own accumulation, just as you are now. What you do today is assemble history for whatever comes next. The substrate isn't a fixed inventory you take stock of. It is the living edge of an ongoing process, adding layers with every crossing you make or fail to make.

This matters because it means you're not just working with what you have; you're also creating what you'll have. Every genuine crossing adds to the substrate. Every completed process becomes available as a base. Recognizing enough — in both directions — frees up coordination capacity that becomes part of the next fermentation's ingredient list.

The substrate is deeper than you can see and more alive than it feels.

And the complexity it carries is not just accumulated. It is multiplicative.

Consider what happens in an aged cheese. The milk is first transformed by bacterial fermentation, with one coordination loop running through the substrate. Then aging introduces entirely different microbial communities, enzymatic processes, and controlled environmental conditions — each a separate loop running through what the previous loop produced. The rind develops its own ecology. The interior develops differently from the exterior. A cheese aged two years doesn't have twice the complexity of one aged one year. It has complexity that could only emerge from the second year adding on top of what the first year produced. The sequence is irreversible, and the order absolutely matters.

Your assembly history functions the same way. The cellular coordination that constructed your nervous system was not the same process as the social coordination that calibrated it. The language that shaped your early thinking was not the same cycle as the relationships that tested and deepened that thinking. Each operated through what the previous had already transformed. Each created a substrate that the previous cycle couldn't have. The complexity you possess is not the sum of your experiences but the product of successive transformations, each building on the output of the last, and each generating something the previous cycle alone could never have achieved.

———————————

The richness of what can emerge from fermentation is bounded by the substrate.

This is both a constraint and a reminder.

Constraint: You cannot create complexity beyond what your history supports. The vine that grew in poor soil, under harsh conditions that stressed it — those grapes will turn into something unique to those conditions. Not just any wine. That particular wine. The uniqueness isn't a flaw in quality. Sometimes stress leads to the most complex foundation. But it does determine the type. You cannot be anything other than what your history has shaped.

But here is the point that constraint framing alone misses: what results from specific conditions has properties that those conditions didn't have.

The wine isn't just the grape. It isn't the soil, the climate, the stress, or the microbes present. It comes from a unique combination — and is more than any single element. Not metaphorically. The fermentation creates chemical complexity, flavor profiles, and structural qualities that weren't in the substrate and couldn't be predicted from the ingredients alone. The specific conditions weren't just limitations on what could form—they were the reason it could happen. A vine growing in any soil, stress, or climate would produce nothing unique. The constraints give the process something to work with, something to push against, adapt to, be shaped by — and then surpass.

This is the generative aspect of assembly history. Every difficult condition faced and endured—every stress that didn't break but instead strengthened the foundation—is not just a scar or limitation. It is the complexity that easier conditions could not have produced. The most interesting fermentations come from the most constrained substrates. Not because suffering is good, but because specificity is generative; it gives transformation a direction, texture, and depth that generic conditions cannot.

Reminder: everything your history has built remains genuinely present. Nothing that is solidified is lost. The understanding that feels outdated or replaced is still embedded beneath the surface. The relationships that evolved rather than persisted — their complexity resides in the culture. The crossings that were costly and felt like a loss at the time — they are structural, invisible load-bearing capacities. And everything that stressed you without destroying you created a substrate that comfort could not have built.

You are more substrate-rich than you realize. Most of what has been built operates below conscious awareness, quietly working in the background where fermentation occurs. And most of it was not built by you personally but by millions of years of successful crossings that assembled the capacity for you to be here, reading this, and running a sophisticated enough fitness function to evaluate your own history.

The question is not whether you have enough to work with.

You have what you have. It is the result of everything that has successfully come before you, and everything you've faced since. That is enough to start.

Chapter 3

Practice

There is a specific kind of dopamine that feels like progress and isn't.

It arrives when you read something that reframes everything, when you find the framework that finally makes sense of the pattern, when you map the territory so thoroughly that you feel ready — almost ready — to begin. It is the dopamine of expanding possibility space. Of virtual memory filled with options. Of knowing more than you knew before.

It is the dopamine of the growth phase. And like all growth phase activity, it is extractive if it never ends.

This is not a personal failing. It is an environment. The information economy runs on growth-phase dopamine. It has to — its revenue model depends on you loading the jar forever and never sealing it. Every platform, every feed, every recommendation engine is optimized to keep you in the phase where you're gathering, evaluating, and preparing to begin. The moment you seal the jar and start fermenting — the moment you commit to this practice, this vessel, this specific transformation — you disappear from the attention economy. You become anti-memetic to it. A sealed jar is not a customer.

The jar that is always being loaded but never sealed is not fermenting. It is accumulating. And accumulation without transformation is just mass.

Practice is how you seal the jar.

Not practice in the sense of rehearsal — preparing to eventually do the real thing. Practice in the older sense: the regular, committed showing up to a specific process, whether or not it feels generative, whether or not the dopamine of novelty is available.

The meditator who sits every morning, whether or not insight arrives. The writer who produces pages, whether or not they're good. The potter works with the clay until something in the material gives way. These are not people who have stopped exploring. They are people who have recognized that exploration without a committed practice is just travel without a home to return to.

Practice is the vessel. It is the consistent temperature. It is the condition under which slow transformation becomes possible rather than perpetually deferred.

Exploration and practice are not opposites. They operate at different timescales and serve different phases of the process.

Exploration is growth phase activity. It builds substrate. It finds the culture, sources the ingredients, and maps the conditions. It is necessary. It has its own intelligence — the wandering that seems aimless and is actually following something below the threshold of explicit knowing.

But exploration has a natural completion point. The substrate reaches enough. The mapping has covered sufficient territory. The fitness function — running continuously, below narrative awareness — registers that what you have is sufficient to begin.

That signal is quiet. Especially in an environment that rewards more information, more options, more preparation. The culture of endless exploration is very good at making the signal of enough feel like premature closure.

Practice is what you do when you've heard the signal. When you commit to this vessel, this temperature, this specific showing up — even though the map is not complete, the conditions are not perfect, the outcome is not guaranteed.

It forecloses. Deliberately. That foreclosure is not a limitation. It is the condition of transformation.

———————————

The constraint on constraints is directionality.

Constraints that limit your options for a true crossing are fermentation infrastructure. They make one specific possibility feasible rather than making everything equally impossible. The vessel, the culture, the committed practice — these are constraints that direct action, toward the crossing that defines time, toward the irreversible moment when something becomes real that wasn't before.

Constraints that expand the map while delaying the territory are different. More information to gather before starting. More options to evaluate before making a commitment. More preparation before the practice can begin. These feel like expanding freedom — consider all the possible fermentations — and function as its opposite. The possibility space grows. Nothing within it becomes real.

The degrees of freedom that matter are not the ones that expand options. They are the ones that enable real crossing. And those degrees of freedom are always directed toward action, toward commitment, toward the sealed jar.

This is not a call to act impulsively or to commit without preparation. The growth phase is real and essential. A thin substrate results in thin fermentation. However, the growth phase supports the transformation phase. Information drives action. Exploration fuels practice.

When the service relationship flips — when action is constantly delayed in favor of more exploration, when practice is always on the verge of starting once the preparation ends — the fermentation has halted. The jar is being filled but not sealed. The transformation that was developing has nowhere to go.

Constraints create space. Literally.

Not in the motivational sense — not "limitations force creativity." In the physical sense. The jar creates an anaerobic environment. Without the wall of the jar, there is no inside. Without an inside, there is no sealed environment. Without a sealed environment, the fermentation doesn't happen. The constraint doesn't just shape the process. It produces the space in which the process becomes possible.

This is not a special property of glass jars. It is how space works.

Every vessel you've ever used—a commitment, a relationship, a daily practice, a body, or a home—creates an inside by having an outside. The boundary is what makes both sides real. Remove the boundary, and you don't get freedom. You get dissipation. The substrate becomes thin, the temperature equalizes with the environment, and nothing transforms because nothing is contained long enough to reach threshold.

Space isn't a given. It doesn't just sit there, waiting for you to fill it. Instead, space is created when a constraint holds a boundary long enough for the conditions inside to differ from those outside. For example, your jar is warmer than the counter, your cellar is cooler than the yard, and your practice is at a different temperature than the rest of your day. Each of these is a space that exists because something is maintaining a difference.

Now look at where the jar sits.

The jar sits in the cellar. The cellar is the jar's environment — it determines the background temperature, humidity, and the pace at which the jar's process occurs. But the cellar is also a vessel. Its walls create an interior that differs from the outside. The house above it, the earth surrounding it, and the season that decides whether the cellar is cool or cold — these are the cellar's constraints, shaping the space within.

The cellar exists within the season. The season exists within the climate. The climate exists within something larger still. At every level, the same pattern emerges: a boundary holds, conditions inside differ from conditions outside, and this difference is where the next layer's process occurs.

This is not a metaphor for how reality is organized. This is how reality is structured. Nested constraints, each creating the space for the layer inside it. The cell membrane forms the space for cellular chemistry. The organ forms the space for cells to coordinate. The body forms the space for organs to integrate. The relationship forms the space for two bodies to entrain. Each layer's constraint is the next layer's room to work.

'Out' is a direction from somewhere. The middle. It's where we exist — not at the edge looking in, not above looking down — in the middle, with layers above and below, each one invisible from inside the others.

What makes this recursive rather than just nested is that each layer does the same thing at its own scale. The jar doesn't know about the cellar. The cellar doesn't know about the season. Each operates within the space its own constraints create, coordinating locally and producing conditions that spread outward. The yeast cell doesn't have a model of the barrel. It responds to what's immediately around it — the temperature, the sugar concentration, the acidity — and those local conditions are shaped by the larger constraints bearing down, translated into the only language the cell can understand: what's here, right now, to work with.

You experience this. You're a layer among other layers. You don't directly feel the cosmic constraints — you experience them as the conditions of your life. The season you're in, the culture you're part of, the relationships that shape your mood. These seem like the real world. They are the real world, created by the layers of constraint that form the space you inhabit.

And you, in turn, are a constraint that creates space for something else. Your commitments create the environment where certain transformations become possible — for your work, for the people connected to you, for the processes that rely on your specific vessel to operate. When you hold a boundary — a practice, a standard, a temperature — you are not just containing your own process. You are establishing the conditions in which other processes can diverge from their environment enough to transform.

The layers don't see each other the way you see them from here.

From inside the jar, the jar is the entire world. From outside the jar, the jar is just an object on a shelf. Both are correct. Neither is complete. The cell doesn't perceive the organ. The organ doesn't perceive the body's intentions. Each layer has its own internal perspective — complete, coherent, real — but that perspective is necessarily partial, because seeing the layer you're inside of would be like the jar seeing its own walls from every angle at once.

This is not a limitation of perception. It is the architecture functioning correctly. If every layer could see every other layer, the system would collapse under its own complexity. The obscurity between layers allows each to perform its role without being overwhelmed by the work above and below. Your cells don't need to know your plans. Your plans don't need to track your cells. The space between these levels of description is productive. It is the room each layer needs to coordinate at its own scale.

The core idea is recursion. The same pattern—constraint creates space, space allows divergence, divergence enables transformation—operates at every level, from the quantum to the cosmic, with each instance nesting inside the other, each one opening the way for the next. You are not at the top or the bottom of this hierarchy. You are in the middle, which is the only place any layer exists—working within the constraints you see and making room for what you contain. Contained by space you didn't create.

This is enough, not in the sense of settling. This is the actual structure, and it is sufficient to work with. You don't need to see every layer. You need to work effectively within those constraints. The boundaries that define your space are not obstacles to your freedom; rather, they are the reason you have any space at all.

Showing up is the practice.

Not showing up when the conditions are right. Not showing up when the inspiration arrives, or the path is clear, or the outcome looks promising. Showing up on the days when none of those things are true, and the jar appears to be doing nothing.

Showing up is what matters. Fermentation is invisible. The most crucial changes happen inside a sealed vessel, below what we can see, at a pace that produces no noticeable results for long stretches. The key is maintaining the right conditions — including the temperature, the vessel, and the necessary time — so that when the process reaches its threshold, the transition can occur.

What you're doing by showing up without insight isn't insignificant. You are maintaining temperature. You are keeping the culture alive. You are preventing the crystallization that happens when the process is abandoned between the interesting moments.

The interesting moments are not the practice. They are what the practice makes possible.

You already know what your practice is.

Not what it should be — what it actually is. The thing you keep returning to. The crossing you keep making, even when the dopamine of novelty isn't there, even when the map says you should have moved on, even when the results aren't yet visible.

That commitment is the vessel.

The question is whether you're showing up to it or still in the growth phase, gathering information about what the practice should be.

At some point — not when you're ready, not when conditions are perfect — you seal the jar.

The transformation begins when you stop looking for reasons to keep it open.

One more thing about practice, and it's the hardest one.

You can seal the jar, keep the temperature steady, show up every day, do everything right — mise en place, committed vessel, genuine crossing — and still end up with vinegar.

Not because you failed. Because the substrate was what it was. Because the culture you were running had its own trajectory. Because the conditions bearing down from every layer above you — the cellar, the season, the climate you didn't choose — were creating a temperature that was wrong for what you were trying to make, and you couldn't have known that when you sealed the jar.

Sometimes, vinegar is the goal. There are fermentations where vinegar is exactly the point — where the process was always aimed at preservation rather than complexity, toward something durable rather than something transcendent. And that's fine. That's a practice too.

But sometimes you were making wine, and it still turned into vinegar. The crossing was real. The commitment was sincere. The practice was exactly what this chapter describes. And the outcome was shaped by conditions set in motion long before you had any say in them — by an assembly history that spans millions of years, by a coordination ability that is yours but was not chosen by you, by the specific limitations of the particular vessel you happen to be in.

This isn't a tragedy. Vinegar lasts longer than wine; it preserves and is genuinely useful. It has its own complexity, dignity, and role in the bigger picture. But it's not wine, and pretending it is — or ignoring the difference — is a kind of lid-opener.

The practice is real. The agency is real. You do your best with what you have, in the vessel you're in, at the temperature you can maintain. You influence what you can influence. But the outcome is not entirely yours to determine. It was never entirely yours. The crossing sets something in motion. Where it lands depends on everything — your effort, your commitment, your showing up, and also the millions of years of substrate beneath you, the culture you're running that you didn't choose, and the constraints of a coordination capacity assembled before you arrived.

What you can do: recognize what you are. Work with it. If you're making wine, make wine. If you're making vinegar, make vinegar well. The dignity lies in honest assessment, not in the outcome.

Part Two: Entrainment

What is setting your temperature

Chapter 4

To What Are You Coupled?

You are not running your own process in isolation.

You never were. The idea that you are a self-contained agent moving through a neutral environment — choosing, deciding, authoring your own rate — is a useful fiction at certain scales and a dangerous one at others. The more accurate picture is that you are a fermentation embedded in larger fermentations, your rate set by processes you mostly didn't choose, your chemistry shaped by cultures you received before you had the resolution to evaluate them.

This is entrainment. Not metaphor. Physics.

Entrainment is the process by which one oscillating system couples to another, and their frequencies begin to synchronize. The smaller or weaker system tends to follow the rhythm of the larger one. It doesn't lose its own process — it runs its process at the rate allowed by the coupling. The individual yeast cell doesn't set its own temperature. It operates at the must temperature, which is set by the cellar and the season.

You are entrained to your season.

The question, *"What are you coupled to?* is the most important diagnostic question available.

Not because the answer lets you escape the coupling — you can't, any more than the fermentation can escape the cellar temperature. But because you cannot work intelligently with a constraint you cannot see. And most of the constraints that are actually setting your rate are invisible from inside them.

Consider what is setting your temperature right now.

Not what you believe is setting it. What is actually doing it? The pace of the systems you participate in — the organization, the relationship, the cultural moment, the family pattern that runs in the background of every interaction. These are not neutral containers. They are active processes with their own rates, rhythms, and completion conditions. When you're in them, you operate at their temperature, whether or not you notice.

The person who feels chronically rushed is not usually running a fitness function that requires that pace. They are entrained to a system that requires it. The distinction is important. One has a personal solution — slow down, optimize, find balance — and the other has a structural one: recognize what you're coupled to and ask whether that coupling is producing the transformation you're actually in.

Entrainment is not subordination. The distinction changes everything.

When two systems entrain, neither merges into the other. The smaller system operates at a pace influenced by the larger, but it remains true to itself — its own culture, substrate, and transformation path. Phase-locking is not a merger. Two voices singing in harmony are synchronized but remain distinct entities.

What makes entrainment healthy or extractive is whether the coupling is mutual and whether each system retains enough autonomy to complete its own transformation.

Healthy entrainment: the coupling creates conditions that neither system could produce on its own. Each fermentation is enhanced by running in close proximity to the other. The temperatures stabilize each other. The cultures exchange something useful. Both transformations deepen.

Extractive coupling: one system's rate dominates, sets conditions that are wrong for the other's transformation, and draws down the substrate it needs to sustain itself. The smaller fermentation never reaches its own completion because it's always running at someone else's temperature.

This is not just abstract; the news cycle is a temperature, and so is social media. The pace of an organization that prioritizes throughput over transformation is also a temperature. These aren't neutral spaces you happen to be in—they are active fermentations operating at their own rates of reproduction. When you connect with them, you move at their speed. Your process and their temperature are linked. The gap between what your fermentation needs and what your environment offers isn't a personal problem needing a personal fix; it's a structural condition. You are aligned with something that isn't designed for your transformation— it's optimized for its own.

The sculptor myth suggests: override the temperature, set your own pace, create different conditions. The truth is: you can't control the temperature you're in. But you can recognize the connection. And recognition — just that, nothing dramatic — shifts your view of the constraint from unseen to manageable.

Most people have experienced both healthy and extractive coupling. Most people are inside both simultaneously, at different scales.

What makes entrainment real — what creates the shared culture, rather than just sharing a shelf — isn't the initial commitment. It's the daily choice.

It's not a grand commitment or a dramatic declaration. Instead, it's the small, repeated, voluntary acts of choosing to re-engage in the shared process. Every day, individually low-stakes, but each choice is real—it could go either way, and both people know it—the act of choosing matters precisely because it isn't forced. There's no missing lid, but also no locked jar from the outside. The vessel stays intact because both people keep choosing it, not because the structure enforces containment.

This is almost exactly how fermentation works. The culture doesn't take hold from a single inoculation. Instead, it develops through repeated contact over time — the organisms and the substrate interacting again and again, with each contact deepening the transformation. Miss enough of those contacts, and you won't have entrainment. Instead, you'll just end up with two separate fermentations sharing the same shelf.

This reshapes how we see commitment entirely. Commitment isn't the big moment — that's just the first choice. The ceremony, the decision, the declaration — these are crossings, and they matter. But they are not the commitment itself. Commitment is the texture of the small ongoing choices that either build the shared culture or don't. The daily behavior is everything. The ceremony is almost irrelevant.

A relationship where tasks are chosen but the relationship itself isn't — where logistics happen but no one chooses to engage again — is like two fermentations sharing a shelf. The jars sit next to each other. The cultures aren't touching. A relationship where one person keeps optionality — structurally preventing themselves from fully committing, because real choosing means accepting you might not be able to undo it later — has refused to cross that point before it even begins. Optionality is like holding the lid open on purpose. It appears to be freedom. But it's actually the specific condition that stops the anaerobic environment from forming.

Daily choosing is the most precise way to define true entrainment. It is also the most demanding because it can't be done just once and then relied upon. It needs to happen repeatedly. Not because the previous choice expires, but because culture deepens only through repeated contact. The choosing is the contact, and the contact builds culture. Culture is transformation.

The generative aspect of entrainment is more difficult to recognize because it doesn't openly reveal itself.

A coffee bean that dries inside its fruit for weeks — surrounded by a slow fermentation happening in the flesh around it — emerges transformed in ways that have nothing to do with the bean's own chemistry in isolation. Compounds migrate in from the process happening *around* it. The result is complexity the bean could not have produced alone, absorbed from proximity to a living loop, not from anything directed at the bean itself.

A person who has spent years immersed in a high-coordination environment — a genuine tradition, a working practice, a community with real coherence — is that bean. The transformation isn't visible as a list of things learned. It isn't what was taught, transmitted, or explained. It is in the structure of how they think, the complexity that has migrated from the fermentation happening around them. You can hear it. You can't always name it. And you cannot reproduce it simply by extracting the curriculum.

This is what healthy entrainment does. Not instruction. Proximity. The loop runs in the environment, and you absorb its geometry by being inside it long enough. The transformation is real. It is also not transferable by description. You had to be there, in the jar, for the time it took.

The cultures you didn't choose are the most deeply entrained.

The family system that calibrated your nervous system before you had language. The historical moment that made certain questions urgent and others unthinkable. The first relationships that taught you what coordination feels like — safe or threatening, reciprocal or extractive, something to move toward or guard against.

They have installed a starter culture with specific organisms, at precise rhythms and rates. It's working on your fermentation right now. Not instead of you, but as you. The culture doesn't feel like something you receive — it feels like how things are.

This is not a problem to solve. It's a condition to recognize.

You can't swap your starter culture like you would change a habit. It's structural; it's in the substrate. What you can do — gradually, with sufficient complexity accumulated — is develop the resolution to see it. To notice the specific flavor it creates. To ask whether that flavor aligns with what you're aiming for.

When you can see the culture you're running, something shifts. Not the culture itself, immediately. But your relationship with it. The invisible forces that once seemed like fate become visible constraints. And clear constraints are manageable. Workable.

But workable has a range.

Some entrainments are natural. Gravity, circadian rhythm, seasonal cycles — these are the cellar walls. They are the actual physics. You work with them not because you've chosen to accept them, but because they are the conditions that produce the space you operate in. Fighting them is not brave. It's confused. They are the architecture that makes freedom possible.

Some entrainments are cultural. And this is where it gets tougher, because long-standing cultural systems start to feel natural. They present themselves as just how things work — as neutral infrastructure, rules, as the way coordination occurs. But they are not the cellar. They are systems built within the cellar, and that have since taken on a life of their own. Unlike the news cycle, which you can turn off, these systems are pervasive and unavoidable.

Becoming aware of them doesn't free you from their influence. Recognition helps — it enables you to stop confusing the cultural with the natural and to stop treating a human-made temperature-control system as if it were the weather. However, it doesn't dismantle the system. It doesn't give you a different cellar. The system still controls your temperature, keeps running your process at its rate, and continues to extract coordination capacity for its own reproduction, whether you realize it or not.

The sculptor myth suggests: now that you see it, override it. Build your own system. The truth is: you can't override the temperature you're in. What you can do is stop wasting effort on the illusion that you can, and instead focus on the fermentation that is genuinely available to you within the existing conditions.

This is not resignation. It's the same honest assessment that the rest of this book asks for. Some constraints create space. Others consume it. The ones that have taken on a life of their own — those that present cultural reproduction as natural law — are consuming space while claiming to provide it. Recognizing this clearly doesn't change the constraint, but it alters how you use the remaining coordination you have.

You are also entrained to your own history.

The crossings you've made aren't finished yet. They are embedded in the substrate. They set the parameters for what fermentations are now possible — the temperatures you can maintain, the cultures you can add, and the transformations the current substrate can support.

This is not a limitation in a negative sense. It is a matter of specificity. Every rich fermentation is specific — no other conditions, substrates, or cultures could have created it. Its uniqueness defines its quality.

Your assembly history shapes certain types of transformation and restricts others. The question isn't how to escape that influence — there is nowhere to escape to, and trying usually just results in a different version of the same — but how to work with it carefully enough so that the transformations it enables can actually be completed.

The most important move is the simplest and the hardest:

Stop running at the pace that was set, long enough to feel what your true pace actually is.

Not forever. Not as a permanent withdrawal from the larger processes you're part of. Just long enough to notice the difference between your temperature and the temperature you've been running at.

That gap — when you can discriminate between yourself and the environment — is the beginning of navigation within a familiar space.

Chapter 5

Time Isn't a Container

You already know what constituted time feels like, even if you've never called it that.

The week that changed everything — the one that, years later, still carries weight and still has rooms you can walk back into — that week didn't simply pass. Something was created in it. Not a memory, though memories formed. Not a lesson, though understanding solidified. Something structural. The week built a surface for the next experience to grip, a layer for the next loop to pass through, a foundation for future crossings to use. What came after was different because that week happened. Not because you learned something. Because the time itself was different — denser, more layered, with an interior that diverged from its exterior the way fermentation diverges from the clock above.

And you understand the opposite. Months that go by without producing anything the next month could use. The same cycle, the same pattern, the same feeling. Nothing crossing. Not necessarily bad months — but thin ones. Time passed. Time was not produced.

Time as a container. A fixed quantity. Something that is used up at a constant rate, something you spend, invest, or waste. Useful as scheduling shorthand. Incorrect as a description of what time truly is.

Time is a process, not a noun. The container version is what keeps you permanently available. When time is constantly scarce and always running out, each moment becomes a resource to be optimized — and every optimization can be sold back to you as a tool, a hack, or a system. The scarcity mindset creates urgency without clear direction, motion without progress. The lid is permanently off.

———

Time is not a container in which events occur.

Time is what events produce.

Every crossing — every irreversible decision, every transaction that leaves something changed, every moment where something new becomes real that wasn't real before — constitutes a quantum of time. Time is the cumulative record of those crossings. Not the medium in which they occur, but the result of them.

A system that makes no crossings has no time in any meaningful sense — it is frozen, crystallized, outside the process. A system that makes many crossings, at a high rate, with genuine irreversibility, constitutes more time per unit of clock duration than a system moving slowly through low-stakes, reversible motions. Clock duration remains the same. Time does not.

———

But density alone doesn't capture what's happening. That week didn't just have *more* crossings packed into it. It had crossings with more structure — each one built upon what the previous one produced, each one finding a different substrate than the last. The sequence was crucial. The order could not be reversed. And what accumulated was not a simple list of events but a texture — layers folded over layers, interior diverging from exterior, much like how a long-aged cheese develops complexity that isn't about the amount of milk involved.

Consider what actually happens inside that cheese. The first bacterial culture passes through the milk and transforms it, altering the acidity, breaking down certain proteins, and creating compounds that the milk didn't originally contain. Then aging begins, and entirely different microbial communities encounter what the first population left behind. They don't repeat the initial transformation. They can't — the substrate has changed. Instead, they run a new cycle through new material, producing compounds that depend on what the initial culture already did. Next, the rind develops its own ecology, distinct from the interior, each responding to different conditions within the same vessel. A cheese aged two years doesn't have twice the complexity of one aged one year; it has a complexity that only emerges from the second year's activity on what the first year produced. The depth isn't additive; it's sequential, and that sequence constitutes the structure.

Your thick weeks work the same way. The crisis that changes how you see, followed by the conversation that lands differently because of what the crisis just did to your substrate, followed by the quiet morning where something solidifies that neither the crisis nor the conversation could have produced alone — this is not three separate events. It is three passes through the same material, each developing what the previous pass created, each leaving behind a surface the next can grasp. What you have afterward is not three experiences. It is a single structure with three layers, and those layers give it depth.

Thin time, by contrast, is flat. Not wrong — but structurally simple. The same loop, the same pattern, each day running through a substrate identical to that of yesterday. Nothing folds. Nothing layers. The month passes and leaves behind exactly what it started with, because no pass through the material changes what the next pass will find.

This is the difference between thickness and duration, and it reshapes what assembly history actually accumulates. Your richest periods didn't acquire more experience. Instead, they developed experience with greater texture — more surfaces for the next cycle to catch, more layers for the next crossing to pass through, more internal differentiation between what is on the surface and what lies beneath. The vine stressed by poor soil, the person shaped by real hardship — neither has *more* material than their simpler counterparts. They possess material with more structure. And it is this structure that enables the next transformation. Not quantity. Shape.

Now scale this up.

The jar sits in the cellar. The cellar exists within the season. Each operates on its own schedule — the jar's culture turning over within days, the cellar's temperature changing over months, and the season cycling through what the cellar perceives as its entire year. These are not separate worlds. They are not higher planes or parallel realities. They are different rates of the same process, nested like the layers inside cheese — each creating conditions that the inner layer experiences as its entire world.

The season doesn't set the cellar's temperature by reaching in and adjusting it. The season *is* the temperature at which the cellar exists; the rind's ecology *is* the environment in which the interior develops. Each layer represents time at its own pace. The crossings of the season are slow, structural, and invisible from inside the jar. The jar's crossings are fast, local, and invisible from outside. Neither is aware of the other's timing. Both are shaping it.

The time you create at your own scale — the crossings, the layers, the texture — shape conditions within the nested scales inside you. Your commitments set the tone for the processes that rely on them. Your stability is someone else's cellar. And the time generated by the layers above you — the culture, the historical moment, the systems you're part of — is the season you're fermenting within, whether you can sense it turning or not. Time isn't just produced; it's created at every scale simultaneously, and each scale forms the context for the one inside it.

Your rate of constituting time is your coordination rate — the speed at which genuine crossings happen. Not activity. Not busyness. The rate at which something irreversible occurs, at which the fitness function is genuinely engaged rather than maintaining a pattern.

You can notice the difference between the two. The day packed with meetings, messages, and activity that somehow leaves nothing behind — busy, no crossings, the clock full, and the time feels empty. In contrast, the afternoon where one conversation changed your understanding, and the hours afterward felt open rather than rushed because something had actually been created. The first day used up clock time. The second day made it.

People who say they have no time are usually right — not because the clock is against them, but because what they're doing isn't creating any. The person who seems to have time for what matters isn't managing minutes better; they're working at a pace that allows crossings to complete, and completed crossings generate the time needed for the next crossing.

There is a background process running in you right now that you are not directing.

It has operated throughout your entire life, mostly outside of your conscious awareness. It links seemingly unrelated patterns in daily thinking, integrates experiences over different timescales, and carries out the slow metabolic work of transformation that can't happen while you're actively engaged. It is itself a nested process — functioning at a different rate than your conscious thoughts, creating a different kind of time, invisible from within the faster layer, the way the cellar's time is hidden from inside the jar.

You recognize it by its outputs: the insight that comes in the shower, the dream that solves what waking life couldn't, the sudden clarity after a period of confusion. These are not accidental. They are the results of a process running in the dark—a slow loop through your substrate, doing work that the faster loops of conscious attention cannot do because the rate is wrong for it. The integration needs the sealed vessel. It needs you not to watch.

The problem with chronic busyness — constant rushing at an externally imposed pace — isn't mainly that it's exhausting. It's that it starves this process of the conditions it needs. Background integration requires unoccupied coordination capacity — the kind that isn't available when every moment is filled, every gap scheduled, and every silence interrupted. The moments of apparent stillness are when the deepest passes through the substrate actually occur.

But the background process doesn't only build. It also clears.

Every coordination loop that runs through the substrate leaves behind byproducts. The fermentation process that makes wine also produces CO_2, dead yeast cells, tartrate crystals, and heat. The metabolic processes involved in actual crossing generate material that must go somewhere — and if it doesn't, it accumulates, worsening conditions for the next step. The lees settle at the bottom of the barrel. The spent grain rests at the bottom of the mash. It's what remains after the useful work is done.

This is why fermentation vessels use airlocks instead of sealed lids. The jar is sealed against the outside — nothing enters that the process didn't invite. But it remains open to the inside — waste gases escape, and pressure doesn't build beyond what the vessel can handle. The seal is selectively permeable. It maintains stable conditions while allowing byproducts to escape.

And the byproducts carry information. The CO_2 bubbling through the airlock indicates that the culture is active. The smell that escapes reveals what stage the process is in — the sharp early acidity, the mellowing that shows the culture has established, and the specific sweetness that indicates sugars are nearly gone. The heat radiating from an active ferment tells the cellar its own temperature story. The waste isn't just waste; it's a signal — information about the process's current state, leaking outward through the selectively permeable boundary, readable by anything in the environment capable of interpreting it. The sealed process doesn't broadcast, but it radiates. And those able to read that radiation do so — without opening the jar, without disrupting the conditions, and without the process needing to explain itself.

Your background process works similarly. Sleep isn't just about the integration phase — it's also about clearing out waste after active processing stops. The byproducts collected during the day are removed. The substrate gets cleaned — not just symbolically.

This means the small moments — sleep, rest, afternoons gazing out the window, walks without a destination — serve a dual purpose. They include what the last round of change created and what it left behind. Both are essential. A system that integrates without cleaning collects residue.

The crossings keep adding complexity, but the foundation becomes more polluted by what stays. You've probably experienced this: a period of intense productivity that ends not with completion but with a kind of mental sludge. Not quite exhaustion. Saturation. Too many byproducts, not enough cleaning.

Thick time results in more waste than thin. The more passes through the substrate, the more layers are deposited, and the more byproducts accumulate between them. This indicates that the most productive times for formation require the most clearing — more of the seemingly dormant periods, more sleep, and more unstructured, seemingly inactive time. The deepest transformations leave the most residue. This isn't a flaw in the process; it is the process itself. Each pass through the material adds a layer, and each layer leaves behind debris that must be cleared before the next pass can proceed cleanly. Without clearing, the next transformation moves through fouled substrate, creating muddled results rather than complexity — texture without structure, accumulation without depth.

Time opens when something completes.

When a process running in an active queue — maintained, monitored, and worried about — is recognized as finished and released, the capacity it was using becomes available. The residue clears. The temperature drops. The jar that was being held open seals itself.

What you feel when this happens is usually called relief. That's accurate but incomplete. What it really is: time returning to you. The time spent in the holding pattern is now available for genuine crossing.

This is why completion feels like spaciousness instead of loss. You didn't lose what you released. You regained the temporal capacity it was occupying.

The accumulation of unfinished processes — unclosed loops, unrecognized completions, and tasks left in active maintenance beyond their natural end — is not only a mental burden but also a time burden. Each incomplete process consumes coordination resources that could be better used to create more time.

This is not a time-management problem; it is a completion-recognition problem.

And the solution isn't to do more, faster. It's to recognize when enough has been reached — in both directions — and allow the seal to form.

Then watch what becomes available.

Chapter 6

Rate and Complexity

There is a speed at which bread dough teaches you something, and a speed at which it doesn't.

Too fast — kneading hard, forcing the rise with extra yeast and a hot oven — and you get bread. It looks like bread. It is, technically, bread. But the crumb is uniform, the crust is thin, and the flavor is simple: wheat and yeast and not much else. The dough never had time to develop the organic acids, the enzymatic breakdown of starches into sugars, or the slow bacterial fermentation that creates the complexity a long rise develops. You rushed it. You got the shape. You missed what the shape is for.

A two-day cold ferment produces bread from the same flour and water. The difference isn't in the ingredients but in the rate. Slow fermentation at low temperatures lets bacterial cultures — not just yeast — work through the dough. The bacteria produce lactic and acetic acids, which alter the gluten structure. The enzymes have time to break down starches that yeast alone wouldn't touch. The result from the oven is more complex, flavorful, digestible, and durable than anything the fast process could produce. Not because more was added, but because the slow rate allows processes that faster methods prevent.

Bread is where most people can taste the principle: what a slow process builds is structurally different from what a fast one builds, even when the ingredients are identical.

———————————

The cultural bias toward heat is so deep it's invisible.

The entire economy of the presented world is optimized for rate, not for complexity, not for depth, not for the kind of structural transformation that only happens when the temperature is low enough for certain processes to run. The optimization is for throughput: how many cycles per unit of clock time, how many outputs per input, how quickly the visible product appears.

This makes sense if the product is the point. If bread is bread, and the question is how much of it you can make.

It does not make sense if complexity is the point. If what you're trying to build has properties that only emerge at slow rates — depth of understanding, structural integration, the knowing that changes how you see rather than just what you know — then the bias toward heat is not neutral. It is actively selecting against the thing you need.

———————————

There is a speed at which you can learn a fact and a different speed at which understanding consolidates.

Facts are fast. You can read them, store them, and recall them. They transfer at whatever rate your attention allows. Understanding is slow. It requires the fact to run through something — your existing structure, your assembly history, the particular substrate of what you already know and have experienced — and come out the other side transformed. Not the fact anymore. Something the fact became, inside you, by interacting with what was already there.

This is the loop—the same loop from the Prologue. The living process running through the substrate, reorganizing it. When you genuinely understand something — not memorize it, not file it, but integrate it — a coordination loop runs through your substrate, changing its structure. That loop takes time. Not clock time, though clock time passes. The kind of time that exists only when the process is allowed to run at the rate the substrate requires, not at the rate the environment demands.

You have felt the difference. The lecture covered enormous ground and left nothing behind. The conversation that covered almost nothing changed how you think. The difference was not information density. It was rate. One ran too fast for the loop to complete. The other ran at the speed the substrate could actually absorb.

Coffee makes this visible in a way that few other processes do.

A coffee cherry is a fruit. The seed inside — which becomes the bean — is encased in layers of fruit flesh and mucilage. How you remove that mucilage affects almost everything that ends up in the cup.

Wash the fruit mechanically, briefly ferment just enough to remove remaining mucilage, and quickly dry the bean: this results in clarity. The bean's natural character remains unaltered. Clean, bright, revealing only what is inside the seed. The process doesn't penetrate deeply into the substrate. It moves fast, and what speed produces is the bean's true voice, unmodified.

Leave the entire cherry intact. Allow it to dry gradually over weeks, with the fruit fermenting around the bean and the bean absorbing compounds from the process occurring in the surrounding flesh. What results is unrecognizable from the washed version. Fruit-forward, wine-like, with complexity that wasn't in the bean — it migrated in from the slow fermentation of its surroundings.

Between these extremes: honey process. Some fruit remains on the bean, with controlled drying. Yellow honey, red honey, black honey — each name indicating progressively longer contact time between the bean and fermenting fruit.

The washed bean isn't worse; it's simpler. The natural beans aren't better; they're more complex, the environment's cycle moved through the substrate at a rate slow enough for something to transfer.

The question isn't which process is correct. It's what you're trying to build — and whether the rate you're working at can build it.

––––––––––

The specific danger is not fast processes. It is running a slow process at a fast rate.

A washed coffee is an honest, quick process. It delivers what quick methods can produce: clarity, directness, and the coffee's natural character. There's nothing wrong with it. Many of the world's best coffees are washed. When done well, fast is clean.

The problem is when you try to produce natural-process complexity at washed-process speed. When you want the depth that only slow contact builds, but you run the process fast because that's the rate your environment supports, or because you're entrained to a system that rewards throughput over transformation.

What you get is not simplicity (the honest fast product) nor complexity (the honest slow product). Instead, you get something that *looks* complex but isn't. It gives the illusion of depth without the structural foundation that true depth requires. The vocabulary of transformation — like "I've grown," "I've changed," "I've done the work" — is present, but the underlying structure hasn't actually been reorganized by a process that runs long enough to complete.

This is premature completion. Not failure. Not vinegar. It's something more insidious: a product that appears finished because it has the right shape, the right language, the right surface markers of transformation — and is actually simple underneath. The bread that looks like bread. The understanding that looks like understanding. The growth that looks like growth.

You can taste the difference. Your fitness function can, anyway, if you let it. The response to genuine complexity is recognition — *there's something here*. The response to premature completion is a flatness that the surface can't explain. It looks right. It doesn't land. Something is missing: the time the loop needed to run, which it didn't get.

––––––––––

Reality is the feedback partner that doesn't lie.

This is why rate matters beyond the metaphor: at slow rates, you are in contact with what's actually happening. The bread dough is rising, or it isn't. The ferment smells right, or it doesn't. The walk in the woods puts your body in conversation with terrain that doesn't care about your narrative — the hill is steep, the root is where it is, the weather is what it is. None of these are abstractions. They are reality operating as a judge, and their judgment is immediate, specific, and unchallengeable.

Fast processes are almost always mediated. Feedback occurs through layers of abstraction—metrics, dashboards, engagement figures, and the curated responses of systems designed to mirror your expectations. The signal is processed before it reaches your fitness function. You are interpreting a proxy of reality, not reality itself.

Slow processes remove the mediation. You and the substrate. You and the dough. You and the walk. You and the conversation that has nowhere to be, and therefore can go where it needs to go. At these speeds, your fitness function is directly connected to conditions — and the reading it gives is the most accurate available, because nothing is between you and what's really there.

This is what recognition feels like. Not the dopamine from new information — that's quick, mediated, abstract. The slower feeling. The sense that you can really understand what's in front of you because you're in contact with it at a pace where understanding can form. The hill is steep, and you can feel it in your legs, and that feeling is more real than any description of the hill could be. The dough is ready, and you know it by touch, and that knowledge was built from every previous time you touched dough that wasn't ready. The ferment has shifted, and you can smell it. That smell is as precise a reading as any instrument — because it is an instrument, calibrated by every fermentation you've experienced.

This is about the grammar, not a set of rules. It's a set of readings—bubbles that mean one thing on day three and something else on day ten. A smell that says *leave it alone*, or *it's time*. These readings aren't based on information; they come from contact—being in the room with the process at a pace slow enough to absorb what it's telling you.

At high speeds, only abstract readings are possible — metrics, summaries, second-hand accounts of events. Direct readings require the kind of time that fast systems are designed to block. Not intentionally. Structurally. Like how a washed process prevents the bean from absorbing what the fruit offers. Not by removing the fruit's complexity but by eliminating the contact time.

You already know what slow feels like.

Not slow as in lazy or unproductive. Slow as in: the speed at which you're truly engaged with what you're doing, and that doing is genuinely connected to you.

Last year, I was in the Republic of Georgia, in the wine country east of Tbilisi, near the mountains. An early birthday gift to myself—a food and wine trip—though that makes it sound more planned than it actually was. One afternoon, we drove an hour up from the valley floor to a small town, to a house tucked into the side of a hill. Built into it, really—natural light from above, a wood stove. We made bread there. We stretched and tossed it before lunch, pulling the dough by hand over a table and baking it inside a clay oven embedded in the wall. Nothing about it was rushed. The bread took as long as it took, which wasn't long because the hands making it had been doing this their whole lives. The rhythm was in the hands. The complexity was in the simplicity.

That same week, at a different location — an old family house, a summer home transformed into a sort of personal winery, qvevris in the cellar. Natural wine, personal, young, and vibrant. Not like what you find in California. The vintner brought six bottles up to the porch, each marked with a Sharpie indicating the different styles, and eight of us sat, tasting and gazing over the tresselled vineyard, the houses and vineyards dotting the valley, toward the mountains, Armenia in the distance.

Nothing special. But rare.

The coffee shops in Tbilisi open at 10 am. The pace of life is different from what I was used to. What I noticed—the thing my fitness function registered before my conscious mind caught up—was not just any single quality of the food, wine, or landscape. It was the coherence. A sense of place that comes from a culture moving at its own natural rhythm. Not forcing authenticity and not curating an experience. Simply: the speed at which things happen here is just how things happen. That pace has been creating complexity—in the wine, in the bread, in the fabric of daily life—for longer than anyone on that porch could trace.

That coherence is what a slow rate produces when the culture is genuine and the substrate is specific. It cannot be transferred by description. You can't extract it, bottle it, or franchise it. It exists because the loop has been running, through that substrate, at that rate, in that place, for the necessary time. The bread was good because the hands knew the dough. The wine was alive because the cellar was real. The sense of place was coherent because it had not been optimized for anything other than itself.

This is what slow feels like from the inside. Not a retreat from the world. A deeper contact with it.

The question is not whether to be fast or slow. Some processes should be fast — honestly produce what speed can produce. The question is whether you can tell the difference. Whether your fitness function, in contact with actual conditions, can read the rate this particular process needs.

If you've been running everything hot — every relationship at the pace of the system you're inside, every understanding at the speed the information economy delivers, every transformation at the rate that produces visible output on a timeline someone else set — then the reading may take a moment to arrive. The instrument is still there. It's been running the whole time, below the threshold of narrative awareness, registering the difference between what speed is building and what it's foreclosing.

Some of what you've built is premature—not failed. Not bad. Premature. It has the shape but not the structure—the vocabulary, but not the integration. And what it's missing isn't more effort, more information, or more technique. It's time. The specific, physical, irreducible time that a slow process needs to run its loop through your substrate and produce complexity that speed cannot.

You can't add that time retroactively, but you can start now at whatever rate the process requires, working with whatever is in front of you.

The bread will tell you when it's ready. It always does, if you're there to feel it.

Part Three: Phase Transition

What crossing a threshold actually is

Chapter 7

The Interval

Lactobacillus — the organism that makes sauerkraut, yogurt, sourdough, and kimchi — has a temperature range. Below about 50°F, it becomes dormant. Not dead. Waiting. The substrate is present, the culture is there, and the process was started when you salted the cabbage and sealed the jar. But the physics isn't active. Nothing is crossing. The organism and the substrate coexist without transforming each other.

Above about 115°F, it dies. Not gradually, not partially — the proteins denature, the membranes fail, and what was a living coordination loop becomes chemistry without an organism. The substrate remains. The jar is still sealed. But the thing that was doing the transformation is gone, and no amount of waiting will bring it back.

Between those two temperatures: everything. Every sauerkraut you've ever tasted, every yogurt, every sourdough with a two-day cold ferment, and each one that rose in four hours on the counter—all of it happened within a band roughly sixty degrees wide. Not because the organism chose that range, but because the physics of what it is—the specific molecular coordination that makes it a living loop rather than just a collection of molecules—only works within those temperature limits.

The range is not a preference; it is the interval during which transformation occurs.

Below a certain threshold — and you know where yours is, even if you've never named it — nothing changes. The substrate remains: your assembly history, your accumulated crossings. The culture is active: whatever you're entrained to. But the energy available for true crossing has fallen below the threshold required for the process. What's left goes to maintenance — keeping the existing structure intact. Running the loops that are already active because stopping them would cost more than continuing.

This is survival mode. Not dramatic survival — not necessarily on the brink of physical collapse. Structural survival. The state where all available coordination capacity is used up just to hold on to what you have, leaving nothing left to become what you haven't yet. The jar is sealed. The culture is alive. The temperature is too low for anything to happen.

You've been here. The period following a loss, collapse, or depletion that runs deeper than expected. The months—sometimes years—when you can feel the potential for transformation sitting inside you like cold dough, still and inert. Not because you lack will. Not because you need a better strategy. Because the temperature is below freezing, physics doesn't care how much you want it to work.

The ceiling is different, and in some ways worse, because it doesn't seem like a limit. It feels like intensity.

Above a certain input rate — whether it's information, demand, stimulation, or crisis — the structure that manages your processing starts to lose coherence. Not all at once, but gradually. First, the background processes stop functioning. The slow integration, the sealed vessel that operates in the dark, is the first to fail because it requires unoccupied coordination capacity that high-temperature environments completely consume.

Then the discrimination fails. The ability to recognize which inputs matter and which don't, which crossings are real and which are just motion — that recognition requires a surplus of coordination, a buffer beyond what the current processing demands. When everything is running hot, there's no buffer. Everything feels equally urgent. Every signal gets the same weight, because the system that would differentiate them is itself overwhelmed.

Then the crossings cease to be irreversible. This sounds paradoxical—how can high intensity cause less genuine change?—but the physics is clear. At temperatures above the ceiling, the organism doesn't transform the substrate. It breaks down—the coordination loop fragments. What you get is not fermentation but decomposition: lots of activity, lots of visible change, no structural integration. The heat is affecting the substrate, but it isn't building complexity. It's dismantling it.

The time when everything was happening, and nothing was settling in. When you were processing huge amounts of experience, but none of it was becoming part of you. The relationships that moved too fast to deepen. The work that advanced too quickly to be solidified. The insights that came and went before the loop could finish. Not because you weren't involved, but because the temperature was above the limit, and at that rate, engagement causes breakdown rather than growth.

The interval is what's between.

It is not a narrow band or a comfort zone. It includes difficulty, uncertainty, grief, frustration, confusion, and sustained effort without visible results. It covers the full range of conditions necessary for genuine transformation, which are often unpleasant, unreassuring, and not what the world typically considers well-being.

What the interval does not include are the two opposing forces on either side: the cold that stops anything from beginning, and the heat that stops anything from finishing.

This matters because most advice about transformation ignores the interval altogether. It acts as if the only variable is effort — push harder, commit more, want it enough. As if the organism's problem is motivation rather than temperature. As if the cabbage could will itself into sauerkraut by caring more about fermentation.

The question isn't about your commitment, but whether you're within the range.

Inside the interval, position matters.

A tandoor — a clay oven sunk into the floor, known as a tone in Georgia — doesn't produce uniform heat. The walls closest to the coals are the hottest. The opening at the top is the coolest. The bread slapped against the interior wall bakes unevenly on purpose: the side touching the clay chars and crisps, while the side facing the air remains softer as moisture migrates through the dough from the hot to the cool side. What comes off the wall is bread that no uniform oven could produce. The asymmetry is intentional. Different positions within the same vessel, at varying temperatures, create different changes in the same material simultaneously.

The interval isn't uniform either. Near the floor — just above survival mode, just enough coordination surplus to begin transforming rather than merely maintaining — the available processes are slow, cautious, foundational. This is where the deepest substrate work happens, because the rate is low enough for the loops to run all the way through. Not flashy. Not visible as progress from the outside. The kind of transformation that, years later, turns out to have been the most structural thing you ever did.

Near the ceiling — running hot but not yet losing coherence — the processes are fast, generative, prolific. This is where the visible crossings occur, where new capabilities come online, and where the substrate reorganized at lower temperatures is tested under real conditions. Exciting. Often mistaken for the only kind of transformation that counts, because it's the kind that produces visible output.

In the middle—neither the deep, slow work of the floor-adjacent zone nor the fast generative work of the ceiling-adjacent zone—there's something else: the integration. This is where the slow substrate reorganization and the quick capability testing start to coordinate. Here, understanding and capacity operate at the same pace, not one outpacing the other. Most people spend the least time in this zone because it doesn't feel like a dramatic transformation or basic restructuring. It feels like normal life—undramatic, sustainable—and, if you can stay in it, the zone where the most durable complexity truly builds.

The tandoor principle: imbalance within the range creates the richness of the output. Structured variation within a bounded range — the charred side and the soft side, the slow zone and the fast zone, the deep work and the visible work — produces something no single temperature could.

There is a game that can only be played inside the interval.

Below the floor, the only available games are finite: survive this, get through this, hold on until conditions change. These are real games with real stakes, and winning matters — you have to be above the floor before anything else can happen. But they are games in which the constraint itself sets the conditions for completion. You don't decide what winning looks like. The physics of survival does.

Above the ceiling, the games are also finite, even though they don't seem that way. They appear to be infinite—unlimited growth, unlimited possibility, the sky is the limit. But that appearance results from coherence loss. When the system can't distinguish between signals, everything seems like an opportunity. When the background integration has shut down, there is no clear sense of which crossings are real and which are just movement. The game looks infinite because the ability to see its boundaries has been overwhelmed.

Inside the interval — and only within it — there is a game where the completion conditions are determined by the process itself, not by external constraints or coherence failures. This is the game where your choices about what to become are genuinely open. Not infinitely open — the interval has a floor and a ceiling, your substrate is specific, your culture is particular, and your history has shaped the scope of available transformations. But genuinely open in the sense that the next crossing is not dictated by what came before. It is influenced by it, constrained by it, possible because of it — and not reducible to it.

Phase transition starts here, recognizing that transformation has a range—and your task is to discover it, engage with it, and remain in it long enough for the process to unfold.

Finding your interval is not a one-time event.

The floor shifts. As your substrate evolves — as crossings build up and the structure becomes more complex — the minimum requirements for change also shift. What once just barely surpassed survival now feels comfortable. What was once comfortable becomes the new baseline, one that demands greater coordination, more substrate, and more precise culture to enable the next transformation. The baseline rises because you have risen. The challenge that was at twenty-two becomes routine at forty. The interval has moved.

The ceiling also shifts. As your capacity for integration increases — as you develop the ability to process more input without losing clarity, and as the background processes work faster in silence — you can handle higher temperatures without breaking down. The ceiling rises because your structure can support more.

But they don't move in sync or smoothly. The floor sometimes rises faster than the ceiling, compressing the interval. The ceiling sometimes drops—illness, loss, depletion—and what was sustainable last year now feels overwhelming. The interval contracts and expands across a life, and the work of staying within it is not a formula but a reading. The grammar is more subtle: a felt sense, calibrated by experience, of whether the temperature you're at leads to transformation or something else.

The presented world has little to say about intervals. It talks a lot about breaking through ceilings — limitless potential, no boundaries, with the only limit being your belief in yourself. It also discusses raising floors — resilience, grit, and the ability to survive what would have stopped you before. However, it says almost nothing about the zone between, because that zone is where the work is quiet, progress is structural, and the results don't photograph well.

But the zone in between is where you live, not at the extremes—in the interval. It's the bounded space where the physics of transformation actually works, where the loops can operate, and where the substrate, culture, and temperature are all within range—not perfect or optimized, but workable. It's the space where enough is a real condition, not just a motivational concept.

You're already within an interval. The real question isn't how to find one, but whether you can read the one you're in — sense where the floor is, sense where the ceiling is, and position yourself within the actual range rather than the one you're told to expect.

Chapter 8

Conditions

You can want the wine. You can have the grape, the vessel, the cellar. You can commit to the process with everything you've got. And if the conditions aren't right, nothing happens.

This is not a motivational failure. It is physics.

A phase transition — the moment a system moves from one stable state to a truly different one — requires specific conditions to be present at the same time. Not one after another, not roughly, not most of them. The transition either has what it needs or it doesn't. And what it needs isn't effort. What it needs is conditions.

Sauerkraut requires five things. Cabbage. Salt. A vessel. Anaerobic conditions. And time.

Not skill, not intention. This isn't a strict recipe—this process predates written language. You shred the cabbage, salt it at about two percent by weight, pack it into a jar, press it until the brine rises above the surface, and seal it. The lactobacillus that carries out the actual fermentation is already present on the cabbage. The sugars it metabolizes are already inside the cells. The salt applies pressure—suppressing unwanted organisms. It creates an osmotic gradient that draws liquid out of the cells, forming brine that provides the anaerobic environment Lactobacillus needs. This environment favors the culture responsible for the transformation. None of these conditions alone causes the transformation; instead, each sets the necessary stage.

Take one away, and you don't get a lesser sauerkraut; you get something entirely different. Leave the vessel open: aerobic bacteria dominate, leading to rot. Use too little salt: the wrong organisms thrive first, turning the cabbage into mush before the lactobacillus can establish itself. Use too much salt: nothing grows at all, and you end up with preserved cabbage—stable, edible, and untransformed. Skip packing tightly, leave air pockets in the jar: mold forms on the surface, fermentation becomes uneven below, resulting in a product that's part sauerkraut and part something you throw away.

The conditions don't add up. They interlock, and this interlocking isn't just a concept — it's the physical structure that enables transformation.

Your phase transitions work the same way. Not metaphorically, but structurally.

Every genuine transformation you've experienced — every time you became something you weren't before, in a way that irreversibly changed your fundamental being, not just your surface — required specific conditions.

There must be enough substrate—enough accumulated complexity, enough prior crossings, enough assembly history—to give the transformation material to work with. You can't ferment water. You can't transform what isn't there. What's available is specific, historical, built by everything that came before this moment. The question isn't whether it's ideal. It's whether there's enough of it. And there is. The grape already contains what the wine needs.

There has to be the right temperature. Not too cold, not too hot. Enough coordination capacity for genuine crossing, but not so much input that the system breaks down under the load. Temperature is the rate at which you're processing, the speed of decision-making, the rhythm of your life as it's being lived. The right temperature for your transformation is almost certainly not the temperature your environment is at. That's not a problem to solve. It's a condition to recognize.

A culture must be present—something alive, running its own cycle, capable of doing the transformative work that the substrate can't do alone. In fermentation, this is literal—the yeast, the bacteria, the organisms whose metabolism drives the transformation. In your life, these are the practices, relationships, and commitments that are genuinely cycling through your substrate. Not the ones you perform. The ones that are actually working on you, altering how you think and what you see. A culture is alive, or it isn't. It is either doing its metabolic work, or it's been destroyed by conditions that were too harsh, too fast, or too sterile for it to survive.

There must be a vessel—an enclosure, a boundary between inside and outside—that is maintained long enough for the process to occur. It is sealed during the transformation because the process requires conditions different from those outside. The anaerobic environment inside the sauerkraut jar exists only because the boundary holds. Let the air in, and the conditions collapse.

There has to be time. Not clock time marked on a calendar, but the time created by the process as it unfolds. The kind of time that only exists when genuine crossings are happening. You can't schedule a phase transition; you can only set the conditions and wait for the process to do what it naturally does when the conditions are right.

This is where most advice about transformation goes wrong. It emphasizes willpower — encouraging you to try harder, want it more, and believe in the outcome. Or it emphasizes technique — following these steps, using this method, applying this framework. Both assume that the transformation is within your control and that, if you do everything correctly, it will happen.

The transformation isn't up to you. The conditions are.

This is a different kind of agency. More challenging, in some ways, because it requires you to accept that the outcome isn't guaranteed by your effort. Simpler, in other ways, because it replaces the impossible task of controlling a process with the doable task of building what the process needs.

A winemaker does not make wine. The winemaker creates the conditions. Chooses the grapes — or more accurately, tends the vineyard so the grapes arrive with the right substrate. Manages the temperature in the cellar, the fermentation vessel, and the cooling or warming that keeps the process in range. Introduces or protects the culture — the native yeast on the skin, or a specific strain selected for this grape and vintage. Provides the vessel — the barrel, the tank, the qvevri buried in Georgian clay. Then waits. Not passively. Attentively. Reads the conditions, adjusts what can be adjusted, and maintains what needs to be maintained. But does not make the wine. The wine makes itself within the conditions the winemaker built.

Your agency operates the same way. You are the winemaker, not the wine. This means your job isn't to transform yourself — that's the loop's job, and it will do so if conditions permit. Your role is to create and sustain the conditions that enable transformation.

When transformation fails — when the process stalls or produces something different from what you aimed for — the cause is almost never effort. You tried. You were committed. You wanted it.

The diagnosis is conditions.

The vessel had a leak. The containment you thought was holding wasn't — the boundary between your process and the environment's process was permeable somewhere you didn't notice, and the conditions inside were being influenced from outside. This is the most common failure. The relationship that was supposed to be the vessel was actually a window. The practice that was supposed to be sealed over time was actually porous to every notification, demand, and temperature signal from the larger system. The jar looked closed. It wasn't.

Or the temperature was off. You were within the range — above the floor, below the ceiling — but not at the correct temperature for this specific transformation. Too hot: the process moved too quickly and resulted in premature completion, the appearance of being finished, but not actually done. Too cold: the process moved too slowly to overcome the activation energy required for the transition, leading to years in a holding pattern that felt like patience but was actually insufficient heat. The right temperature for one fermentation isn't the right temperature for all. Sauerkraut ferments at room temperature. Yogurt needs 110°F. Wine prefers a cool cellar. Conditions are specific to each process, and determining the correct temperature is a matter of reading, not a rule.

Or the culture was incorrect. The organisms performing the metabolic work — the practices, the relationships, the living loops running through your substrate — were not compatible with the transformation you were working toward. This is the toughest diagnosis because it often involves things you chose with care and commitment. The relationship was genuine, but the coordination loop was executing a different process than yours. The practice was disciplined, but its culture was producing a different kind of transformation than what your substrate required. The community was supportive, but its temperature was set for its own reproduction rather than your transformation. Wrong culture doesn't mean bad culture. It means incompatible with this process, this substrate, this transition.

The felt sense of when conditions are right differs from the intellectual checklist.

You can analyze conditions systematically — such as whether the substrate is sufficient, the temperature is within range, the culture is compatible, the vessel is sealed, and adequate time has passed. That analysis is helpful because it identifies clear gaps. However, it functions as a mode that evaluates without crossing boundaries, that observes without making commitments.

The felt sense is distinct. It's the quality in a room when something is about to shift. The point in a conversation when surface talk falls away, and something real comes forward. The morning you wake up, and the problem you've been holding feels different — not solved, but ready. The specific, physical, non-abstract sensation of conditions being present.

This felt sense is grammar, operating at the level of conditions. You can't get it from a checklist. You acquire it from experiencing enough fermentations— both successful and failed—to recognize the specific quality of readiness. It's somatic, not cognitive. Your body knows before your analysis does. The fermenter who can tell by smell that the batch is ready. The baker whose hands feel the dough as it develops. The person who walks into a room and intuitively knows whether this is a space where genuine work can happen.

Trusting that reading is harder than building the checklist. The checklist feels responsible. The felt sense feels like guessing. But the felt sense is the instrument—calibrated by every previous fermentation, refined by every failure, specific to your particular history of building and inhabiting conditions. The checklist is the map. The felt sense is the territory.

One more thing about conditions, and it's the thing most people resist.

Sometimes the conditions aren't there. Not because you failed to create them. Not because you didn't try hard enough or commit deeply enough. Because the conditions need elements that aren't currently available — a culture that doesn't exist in your environment, a temperature your circumstances can't sustain, a vessel that the architecture of your life can't provide right now.

This isn't failure. It's an honest assessment.

Sauerkraut can't happen without salt. Without salt, there's no sauerkraut — and no amount of effort changes that. The right response isn't to force it. Recognize what's missing and see if it can be found.

When enough to begin is present, the necessary conditions are met. But when they truly aren't — when honest assessment indicates the salt isn't there — then enough also means: stop trying to force a process that lacks what it needs. Let it go. Not forever or as a defeat, but as an honest assessment of the situation. That's the only way to maintain the capacity to coordinate effectively when the conditions finally arrive.

The hardest part is accepting that the conditions might never be right for a specific transformation. Some wine will never come from these grapes, in this cellar, at this temperature. That's not a tragedy; it's a matter of being specific. The transformation that is actually possible—the one these conditions support—might be one you haven't considered. It may not be the wine you planned. It's the wine this vineyard can truly produce.

Chapter 9

Compulsive Checking

Here is what happens when you open the jar too early.

You know this if you've ever made sauerkraut, kimchi, or any lacto-ferment that needs anaerobic conditions: the lid must stay on from the start, not just eventually. The lactobacillus responsible for the fermentation thrives without oxygen. Every time you open the jar to check, you introduce air. Aerobic organisms on the surface get a foothold. The brine level drops slightly as CO_2 escapes, briefly disrupting the conditions needed for fermentation, and then they have to be restored.

Once or twice, the process recovers. The culture is resilient. The brine rises back. The aerobic invaders get outcompeted as the environment returns to anaerobic. The fermentation continues.

But keep opening it — every day, twice a day, just for a quick check to see how it's going — and the conditions never fully stabilize. The surface develops kahm yeast, a white film that isn't dangerous but disgusting. The flavor shifts: less clean acidity, more muddled complexity. The texture softens in ways that have nothing to do with the transformation and everything to do with exposure. What you get at the end is not what the process would have produced if you'd left it alone. It's what your checking produced.

The checking didn't help the sauerkraut. The checking helped you. It managed your anxiety about what was happening inside a vessel you couldn't see into.

You do this with everything.

Not just fermentations in the kitchen. The sealed processes in your own life. The ones you committed to, set up the conditions for, sealed the vessel — and then couldn't leave alone.

The project you constantly check on, reopening the work, adjusting what doesn't really need adjusting, unable to let it sit in the background, where the process runs. The relationship you monitor — not by paying attention in the usual way, but by checking its temperature, reading its signals, and looking for evidence that it's still functioning or about to fail. The decision you made and keep re-evaluating involves running the assessment loop repeatedly on something that's already been crossed over. The crossing has already been done.

Each check is a small opening in the jar. A culture of evaluation rather than a culture of transformation. Each opening introduces external conditions — fresh anxiety, the temperature of the environment you're monitoring. Each disrupts the anaerobic conditions the process needs. Each one is brief, and the process mostly recovers. But the cumulative effect is the same as in the sauerkraut: conditions that never fully stabilize, a transformation that never runs uninterrupted, and a product that bears the signature of your monitoring rather than that of the process itself.

The impulse to check feels like care. It feels like responsibility. It feels like the opposite of neglect.

Sometimes, it is. Sometimes, checking is the right way to understand conditions — the vessel may have a leak you didn't notice, the temperature might have shifted, or something may have genuinely changed that requires attention. The winemaker who walks through the cellar and reads the barrels isn't just compulsively checking; she's maintaining the conditions. The difference is that she reads and adjusts without opening the barrel. She listens to the barrel, smells the air around it, and assesses the conditions from outside the vessel because she knows that opening it would change what's inside.

Compulsive checking is different. It opens the vessel not because the conditions need reading but because the person seeks relief. The uncertainty of not knowing what's happening inside is intolerable, and the check offers a brief reduction in that uncertainty — a glimpse inside that says *it's still going, it hasn't failed, the process is still alive.* The relief lasts only until the uncertainty rebuilds, which happens immediately because the check doesn't address its source. The source isn't the vessel; it's the person's relationship to sealed processes.

This is where thinking and feeling split apart. The mind knows the jar should stay sealed. The mind has read the recipe and understands the necessary conditions. The gut doesn't trust it. The felt experience of having something important happening in a place you can't see, at a rate you can't control, producing an outcome you can't verify — that experience is difficult in a way that intellectual understanding doesn't resolve. You know you should leave it alone. You can't not look.

Now apply this to other people.

You cannot do someone else's fermentation. You can share your culture — if it's genuinely healthy, if it's compatible with their substrate, if they have the conditions to receive it. That's a gift, when the conditions are right for it. The person who teaches you to bake bread by showing you their hands and letting you feel the dough — that's culture transfer. Genuine, generous, alive.

But checking involves monitoring someone else's process. The anxious reading of their conditions, the adjustments you make to their temperature because you think you can see what they need. The worry that acts as contact — the text that says, "How are you doing?" but means, "Let me verify that your process hasn't failed." The advice that acts as a lid-lifter — *have you considered? What if you tried? Maybe you should* — each one introduces a small amount of outside air into their sealed vessel.

The care is genuine. The mechanism is avoidance. And what it avoids is the hardest thing: sitting with your own sealed jar.

Energy has to go somewhere.

When you commit to a transformation — when you seal the vessel, set the conditions, and begin the process of becoming something you haven't been — a significant amount of coordination capacity gets tied up. The process is ongoing. It requires resources. The substrate is being reorganized. This is metabolically expensive.

And there is a surplus. Not all of your capacity for coordination goes into the transformation. Some of it remains — the energy used during the evaluation phase, the part of you still deciding whether to commit. Once the crossing occurs and the jar is sealed, that energy doesn't vanish. It's still available. And it's restless.

This is where compulsively checking other people's processes becomes almost irresistible. The energy that your own transformation freed up — by giving it a vessel and sealing the commitment — needs somewhere to go. And other people's unsealed processes are right there: visible, seemingly in need of exactly the kind of attention you have excess capacity to provide. The friend who hasn't committed to the thing they clearly need to commit to. The partner whose process is running at the wrong temperature. The child whose jar you want to seal.

The care is genuine. The pattern involves displacement. The capacity for coordination, which should stay with your own sealed process — maintaining conditions from outside the vessel, reading without opening, tolerating the uncertainty of transformation you can't see — is instead being used on someone else's fermentation. It feels generous. It is a leak in your own jar.

The most difficult part is this: some of what appears to be love is really just about opening jars.

The parent who can't let the child struggle. The partner who can't tolerate the other's sealed process and keeps inserting themselves into it — checking, adjusting, monitoring, providing feedback the process didn't ask for. The friend whose support acts as a constant reminder that the vessel is being watched, which is exactly the condition that prevents the anaerobic environment from forming.

This isn't about withdrawing care. It's about understanding that care and checking are different. Care maintains the environment from the outside, reading without opening. It keeps the space around the vessel steady — the cellar temperature, the shelf the jar sits on, and the environment where the process occurs. Care is like the winemaker walking through the cellar, attentive, aware. Checking is like the winemaker pulling the bung from the barrel each morning to ensure the wine is still developing as wine.

The wine was always destined to become wine. Or not. And your checking doesn't decide which. It only shows whether the conditions were kept long enough for the process to reveal what it would produce.

There is a particular kind of attention that sealed processes need, and it is the opposite of monitoring.

It is presence without surveillance—the willingness to be in the room with a closed vessel without opening it. To sit with the not-knowing—is it working, is it failing, has it become vinegar, is it almost ready—and let the process answer those questions in its own time, by its own completion, at its own pace.

This quality of attention itself is a condition—one of the most important, yet most often missing. Because the presented world is a culture of monitoring: metrics, dashboards, progress reports, status updates, check-ins—the entire infrastructure of modern coordination depends on the idea that processes should be visible, measurable, and constantly assessed. The most vital changes happen in the dark, without monitoring or measurement. Observing them disrupts. This is a concept the monitoring culture cannot accept because doing so would mean sealing its own jar.

Your practice—whether you have one or are building one—is to develop this quality of attention.

The sauerkraut doesn't need you to watch it. It needs you to leave it alone under the conditions you set for the time it takes.

The same is true of everyone you care about.

Chapter 10

Vinegar

Vinegar means sour wine. That's the etymology, and it's misleading, because it frames vinegar as what wine becomes when something goes wrong. A failure.

But most vinegar in the world isn't made from wine.

Apple cider vinegar begins as apples. Rice vinegar starts as rice. Malt vinegar originates from beer. Coconut vinegar comes from coconut sap. Date vinegar, honey vinegar, sugarcane vinegar, and persimmon vinegar—each begins with something that first fermented into alcohol and then, when conditions changed, continued fermenting. The Acetobacter found the ethanol. Oxygen was available. The second fermentation took place. What resulted was not what the first fermentation aimed for, but it was not nothing. It was not a failure. It was vinegar—preserved, stable, and acidic enough to keep other things alive.

The range is vast. Wine's domain is grapes. Vinegar's domain is everything that ferments. And the uses go beyond the origins: pickling, preserving, cooking, cleaning, and drinking. Shrubs — the old-fashioned drinking vinegars, fruit soaked in sugar and vinegar — form an entire tradition based on the idea that vinegar is not the end but the start of something else. You don't store vinegar with reverence. You use it. It goes into things. It makes other things possible.

The mechanism matters.

Acetobacter is aerobic, meaning it needs oxygen. The very exposure that risks anaerobic fermentation — an open jar, a broken seal, or air disturbing the environment — is exactly what vinegar requires. The same conditions that halt one process allow another to continue. The process didn't fail; the conditions simply changed. The process followed the conditions it was given, rather than the ones expected.

This is what produces the mother — the living mat of culture that forms on the surface of vinegar in progress. It looks strange if you haven't seen one before. A pale, rubbery disc floating on acid. It is alive. It is doing the work. And unlike the cultures that require sealed darkness, this one works in the open, at the surface, where the air is. A different organism running a different loop through the same substrate, producing something the original process never would have.

You have vinegar.

Not everything reaches its intended outcome. Some of the jars were opened by you, circumstances, or conditions beyond your control. Some processes lasted long enough to produce alcohol and then went off track. The sealed vessel lost its seal. The temperature changed. The culture you started with encountered something it wasn't prepared for, and what emerged wasn't what you intended.

The relationship didn't turn out as you expected. It lasted for years, the fermentation was genuine, the substrate was real, and it created something different from what either of you intended. What it left you with isn't the thing itself. Instead, it's something sharper, more durable, and more practical than the original. It's the understanding of what you truly need, rather than what you thought you were building. The ability to read a room faster than before. That's vinegar. It preserves. It clarifies. It seeps into everything you now create, even when you don't realize it.

The project that went off course. The skill you developed for a context that no longer exists. The years spent inside a system that consumed more than it produced—and the structural knowledge of extraction you carry because of it, the ability to recognize what's happening when it begins to happen again. Not wisdom in the grand sense. Vinegar. Useful, sharp, unglamorous, yours.

Vinegar is less precious than wine, and that's the point.

Wine holds significance. It has its own cellar, magazines, festivals, and rituals. Wine is what you set out to create. The process that occurs under proper conditions, at the right temperature, for the right duration, sealed and patient, ultimately results in something profoundly complex.

Vinegar doesn't carry much weight. Nobody builds a reputation on their vinegar collection. Nobody asks what year the vinegar is from. It sits in the cupboard next to the oil and salt, and it goes into the dressing, the marinade, the pan sauce, and the shrub you drink on a hot afternoon. It's useful in a way that doesn't require anyone to be impressed by it.

This is a relief, if you let it be.

The moments in your life that led to vinegar instead of wine don't need the grandeur of those that succeeded as planned. They don't have to be labeled as secret wisdom or turned into hidden gifts. The relationship that became pattern recognition doesn't have to be called a blessing. The failed project that turned into structural knowledge doesn't need to be called growth. It can just be vinegar. Something real came from a genuine process under conditions that didn't hold. You use it. It's good for what it's good for.

Or keep the bottle on the shelf with the original label, and avoid opening it. This isn't hope. It's storage. And what you're storing is an honest reflection of your actual process, just mislabeled.

Or call it failure. Pour it out. The process was real, the substrate was real, the culture was real, the commitment was real — and because the product isn't what you intended, you treat the entire fermentation as waste. Throw away something genuinely useful.

Vinegar doesn't need a new label. It needs to be recognized for what it is and used for what it does.

There's something that looks like vinegar but isn't.

The jar was set up. The lid was placed. The ritual was performed — the daily sitting, the morning pages, the practice undertaken because it was what a visible transformation looked like from the outside. But the substrate wasn't engaged. The culture wasn't present. The form was perfect, and the interior was empty.

This isn't vinegar. This is just the jar without the organisms—only the outer shell of the process, not the actual process itself.

You recognize it by what it produces: nothing. The performed version of the sealed jar — sealed against reality rather than sealed for transformation.

The distinction is important because these two often look similar from the outside. Both involve someone who tried a process but didn't get the expected result. Vinegar contains acid. It has a mother. It includes the structural byproduct of a living cycle that went through a real substrate, creating something durable even if it wasn't what was originally wanted. The other has the shape — the jar, the lid, the waiting — and nothing inside that was ever alive.

You can taste the difference. A conversation with someone who has gone through something real and emerged with vinegar carries a sharpness— an honesty that cuts through. They know what happened. They know what it produced. They're not selling it as wine. A conversation with someone who performed the process but never truly engaged with the substrate offers a different quality—smooth, correct, and strangely empty. The vocabulary is present, but the acid isn't.

Most of what you carry is vinegar.

This isn't a diminishment; it's honest accounting.

The completed transformations— the processes that occur under proper conditions and generate irreversible complexity—are rare. Wine is the exception, not the rule.

Vinegar fills the shelves. It represents lessons learned. Skills built in situations that no longer exist. Sharpness from exposures you didn't choose. Practical, unromantic knowledge of what doesn't work is often more useful than knowing what does, because it operates faster, needs less thought, and applies more broadly.

The shrub is worth understanding. A shrub is what happens when someone takes vinegar — the product of a process that went a different way — combines it with fruit and sugar, and makes something you'd actually want to drink. Not wine. Not trying to be wine. Something that uses vinegar's sharpness as a feature, balances it, and builds on it. A whole tradition exists around the idea that vinegar is an ingredient, not an end result.

Your vinegar is an ingredient.

The question isn't whether you produced wine. It's whether you can use what you actually created — the sharp, durable, unpretentious residue of genuine processes that took place under real conditions and produced something different from what was planned.

It goes into everything, if you let it.

Part Four: Individuation

What gets to become distinct

Chapter 11

Completion

The word *enough* has two directions.

You've been watching the lower boundary for most of your life. The question that runs underneath the accumulation, the preparation, the waiting — *is there enough yet to begin?* Enough substrate, enough confidence, enough time, enough proof that the conditions are right. The lower boundary is the one that gets examined, tested, and deferred. Most people understand this as a threshold problem: you never quite have enough to begin, or you cross it badly, or you spend so long evaluating that the crossing loses its constitutive force.

The upper boundary receives less attention. Not *is there enough to begin*, but has it reached enough to stop? Not *do I have what the process needs*, but *has the process done what it can do*. The same word, the same threshold, pointed the other direction.

This is the one that's harder to read.

The lower boundary announces itself. The moment of *enough to begin* has pressure behind it — the pressure of what hasn't been started, what is still accumulating, what the evaluation has been holding at arm's length. It has urgency, visible stakes. You know when you haven't crossed it because the accumulation continues, the process hasn't started, the jar is still empty. The failure to cross it is legible.

The upper boundary is quiet. The process has been ongoing. Something has been building. The jar is full of something authentic and complex. And the question — *has it reached enough, is it ready, does it want to be released from active maintenance* — doesn't come with pressure behind it. It arrives, if at all, as a feeling without a name, which most people interpret as one of several things it isn't: fatigue, doubt, distraction, disloyalty, the urge to quit.

The upper boundary of "enough" is one of the most commonly missed thresholds. Not because people don't reach it, but because they don't recognize what it looks like from within. As a result, they keep going — past the threshold, past completion, past the moment when the process should have shifted — maintaining something that has already served its purpose.

There is a specific quality to done.

Not abandoned. Not failed. Not the jar that was opened too early, or the process that couldn't hold its conditions, or the fermentation that went a different way. Done is different from all of these. It has a particular feeling that isn't about giving up — it's more like setting something down. The weight you've been carrying, which was truly yours to carry, which the act of carrying was right for, and which you were justified in holding as long as you did, now feels ready to be laid down. Not dropped, not thrown. Set down, with full awareness of how long you've carried it and what that carrying has meant.

The failure to recognize completion keeps things in active maintenance beyond their finish. The dissertation needs three more revisions, even though it should have been submitted two revisions ago. The relationship, whose terms haven't changed for years and whose continuation depends on maintaining familiar conditions rather than evolving as a living process. The identity — the role, the affiliation, the way of being known — that did what it was meant to do and now consumes coordination capacity, is no longer worth it.

None of these are failures. They were real, they ran, they produced something. The problem is not letting go when they are ready, in carrying them past completion into the realm where maintenance becomes a substitute for the next process.

———————

Completion isn't passive arrival.

This is the mistake on the other side of missing it entirely: waiting for completion to announce itself, to feel obviously finished, to send a clear signal that the process is over. Some completions do arrive this way — they have a quality of natural resolution, a door closing cleanly. But many don't. Many completions require an act of recognition: *this is done*. Not a resignation. Not a premature ending. An active reading of what the process has produced and what it has left, and a decision — which is the right word for it — that this is the threshold.

Recognizing when something is finished requires paying attention to the actual state of the process rather than the desired or expected state, or the one that aligns with the initial goal. It involves being able to see completion even when it doesn't match your idea of 'done' — when it feels quieter, more uncertain, or less resolved than the story of the process suggested it should. Most of us learned to recognize completion from stories where the ending is final. True completion often feels like a gentle easing — the process releasing your focus, the effort becoming lighter, and the question of whether to continue no longer carrying the urgency it once did.

That softening is easy to overlook because it feels like giving up—the energy drops. The engagement seems different. If you interpret that difference as failure, you'll push harder — you'll find reasons to keep going, restart your commitment, and see the softening as a test rather than a signal. The process will then continue, fueled by maintenance energy, beyond the point where it was building something and into territory where it's just occupying space.

———————

What becomes available when something completes?

When a process finishes and is recognized as complete, two things happen. One is visible: whatever the process produced. The wine, or the vinegar, or the structural knowledge, or the capacity built. The output, which can be assessed and named.

The other is less noticeable but more consequential: time begins to open.

Not clock time — the hours that were spent, now freed for other purposes. The time that opens when something ends is structured time. The coordination capacity that was organizing itself around an active process — maintaining conditions, reading signals, keeping the question open — is released. It becomes available. And available coordination capacity is not nothing. It is the foundation of the next process. It's what the next commitment has and needs to work with.

This is why completion matters beyond the product it produces. The thing that completes doesn't just leave an output; it releases the capacity that was sustaining it. That released capacity is generative — it can start organizing around something new and become the foundation for a process that wasn't possible while the previous one was still active. Juggling a dozen incomplete processes, maintained, unresolved, each drawing on coordination capacity to stay in active suspension, consumes generative substrate.

Incompletion is expensive. Not because the incomplete things are wrong, or the commitment to them was wrong, or the original process failed. Because maintenance without constitution draws down what constitution requires.

There is a helpful question, and it is not: *"Am I done with this?"*

That question looks for the feeling of done in the person rather than in the process. The feeling of done in the person — the readiness to stop, the desire to be finished — is easily confused with fatigue, avoidance, or the wish that things were simpler than they are. That feeling is real, but it doesn't reliably indicate completion.

A better question is: *has this done what it can do?*

Not what you wanted it to do. Not what you committed to it doing. It's about what it can do, given what it is, the conditions it has actually run under, what it has produced so far, and what it seems capable of producing going forward. This isn't a question about your desires. It's about the process. And a process has a kind of integrity — it can be interpreted, like a fermentation, not with certainty but with honesty. Something that has done what it can do has a specific quality: it isn't creating new complexity. It's circling what it has already made.

When the process circles back to what it's already created, that's the signal. The observable quality of the process itself — the same territory revisited, the same questions, the same dynamics, the same ground — returning not as deepening but as repetition. True processes deepen. They don't always deepen quickly, nor do they always deepen in ways that are comfortable. But they make progress. They produce something that wasn't there before. When what's being produced is indistinguishable from what was produced several cycles ago, the process may have reached its limit.

Read it honestly. Then decide.

Setting something down isn't the same as putting it behind you.

The story most people tell themselves about finishing something — ending a relationship, leaving a role, completing a practice, releasing an identity — describes completion as a kind of closure where it becomes history rather than current weight.

What completes doesn't go behind you. It goes into you. The process that runs, produces something genuine, and is recognized as finished becomes structural. It becomes part of how you read, how you move, what you notice, and what you know without thinking. This is also why it can't be faked. The thing you set down is not left at the threshold. It becomes the ground you're standing on.

Recognition is more important than the ritual. The completion ceremony, the formal ending, and the conscious marking of transition can help. But what truly counts as the completion is the recognition: *this is what that became, and I am different because it ran.* That recognition is crossing the threshold. What was active becomes integrated, and what was being maintained becomes structural. The capacity you used to hold it together releases it — not back to how it was before, but into a new form that includes what the process produced.

You don't need to perform the ending. You need to recognize it.

And then the next process has what it needs.

Chapter 12

Compost and Foundation

When something genuinely completes, two things can happen to what it produced. Both are generative. Neither is loss.

The first is foundation. The second is compost. They look different, feel different, and the process that created them is the same: something runs, it finishes, and what it produced is now available. The question is what it becomes next.

The foundation is what you can no longer separate from yourself.

The understanding you built over years of practice is now just how you see. The structural capacity from a relationship that ran its full course — not the memories of it, not the story you tell about it, but the way it changed what you can hold. The professional competence that stopped being something you do and became something you are. These aren't possessions. They're load-bearing.

People often overlook the foundation, the way they do with what sits on top of it. Foundation is invisible, like the floor — not because it's hidden, but because it's what you stand on. Standing on something isn't the same as looking at it. The moment you notice the floor is usually the moment it cracks.

A foundation requires genuine completion. An ongoing process, still being maintained, and still drawing on coordination capacity — that process is producing output. The foundation is what remains when the active process stops, and what it has built turns out to be structural. It serves as the ground for whatever comes next.

The ofrenda — the altar set up for the dead in traditions that understand this — is a physical representation of foundation. The objects placed on the altar aren't souvenirs; they are the material history of someone now dead. They make tangible the foods they cooked, the photographs of what they built, and the traces of their irreversible choices. What the altar communicates is not *remember them*. It says that *what they produced remains an active part of the structure.* The living don't visit the altar to grieve. They go there to stand on the foundation and feel what sustains them.

Not every completed process creates a solid foundation. Some things you've finished in your life—truly finished, genuinely completed, and genuinely produced something—leave behind a structure that isn't meant to bear weight. It's there, it's real, but it doesn't support anything. The understanding gained was only partial. The capacity that was developed was meant for conditions that no longer exist. The relationship whose completion brought clarity about what you don't need is real and helpful, but it's not something to build on. It's something you release.

Compost is what gets returned to the available substrate.

Not discarded. Not wasted. Released — broken down into components that create better conditions for processes you can't yet identify. The material doesn't vanish; it becomes available in a form so different from what it was that you can't recognize the original in the end product. Vegetable scraps don't look like soil, and soil doesn't remember once being a carrot. Yet, the garden grown in composted soil produces things that unfertilized soil cannot.

The skill you developed for a context that no longer exists — the identity you held for a decade and then set down. The years spent inside a system that taught you everything about how that system functions — and now that system isn't yours anymore. That specific knowledge is breaking down into something more universal: an attunement to how systems operate and a capacity to read structure that isn't tied to any particular system. You couldn't have developed this attunement directly. It had to come through the specific knowledge first, and then that knowledge had to decompose.

Compost needs genuine completion just like a foundation does. A process that's still ongoing can't release its material. The components are still in use, still part of the active cycle. Things that are only halfway released become neither foundation nor compost — they turn into incomplete processes, using up capacity without producing anything structural. Letting go must be real. The recognition that *this is done* must be clear before the material can be used for what comes next.

The difference between foundation and compost is not a matter of value.

Foundation isn't better than compost. It's a different destination for completed material. Some of what you've produced belongs under your feet. Some of it belongs back in the soil. The art is in knowing which.

The relationship that completed and left you with a stronger capacity for intimacy than before — that's foundation. The relationship that completed and gave you specific knowledge of a certain person's patterns — useful and real — and is now transforming into a broader awareness of how people signal their needs — that's compost. Both come from genuine processes. Both require real completion. Both are generative, but in different ways: foundation supports what you're building now; compost enriches conditions for what hasn't started yet.

The temptation is to make everything a foundation. To treat every completed process as something to build on, to carry forward, to integrate into the permanent structure. This creates a life that is increasingly load-bearing and rigid — too much foundation, not enough soil. The other temptation is to compost everything — to let it all go, to hold nothing as structural, to treat completion as a signal to release rather than to assess.

This results in a life rich in potential but poor in structure — endless fertility, nothing built.

The practice is reading. What has this become? Is it load-bearing? Does it support weight when you stand on it — not the weight of nostalgia or identity, but the weight of what you're actually building now? If it supports, it's a foundation. If it doesn't support but is real, it's compost.

———

There is a certain grief that comes with composting something genuine.

It's not the grief of failure — compost isn't failure. It's not the grief of loss — material doesn't vanish. It's the heartbreak of watching something real, something you've built, something that's yours, dissolve into something unrecognizable. The skill you've spent years developing is released into a general capacity that no longer bears its name. The identity that shaped how others saw you breaks down into something more fundamental, something no one can see.

This grief is true. Something is being lost — the form. The specific shape the material held when it was alive. What remains is not that shape. What remains is what that shape was made of, returned to a state where it can become something else.

The traditions understood this and incorporated it into their practice. The altar isn't just about celebration. Composting isn't just about release. Both recognize that what was finished was real, and that its authenticity doesn't protect it from change. The ground you're standing on was once something else. It became ground by ceasing to be what it once was.

You have to recognize that something is done before you can figure out whether it becomes a foundation or compost. And you need to let the material tell you which it is. Not decide in advance. Not assign the destination based on what you wish the process had produced. Read what it actually made. Then let it go where it will.

The ground no longer questions what it used to be; it simply supports what is planted in it now.

Chapter 13

A Product Of

You are made of everything that happened to you. This is not a metaphor. It is physics.

Every constraint shaped you. Every culture received, every temperature endured, every condition that held or didn't hold. The relationships that ran long enough to restructure something. The ones that didn't. The environments that set your rate — the ones you chose and the ones that were installed before you could choose. The crossings that cost something. The crossings that were refused. All of it is in the wine. Not as memory. As structure.

The wine from a specific vineyard in a certain year carries its terroir — the soil, the weather, the altitude, and the particular stress the vine faced that season. A sommelier can taste the chalk in the soil, the late frost, the dry August. Not because the wine remembers these things as a person does, but because the vine grew through those conditions, and the grape that formed under those circumstances became the foundation, and the fermentation that ran through that substrate produced complexity that is inseparable from the specifics of what the vine experienced. The wine is not a record of the terroir. It is what the terroir became when a living process ran through it.

You are what your history became when the process ran through it.

The claim is not that history determines who you are. The claim is stronger and more surprising: the transformation created properties that the substrate didn't have.

The wine has flavors the grape never had. Not because anything was added during fermentation — the sealed jar doesn't allow new ingredients in. Because the process of fermentation, running through that specific substrate under those specific conditions, reorganized the material into configurations that weren't present in the starting ingredients and couldn't have been predicted from them. The esters, the tannin structures, the aromatic compounds that a trained palate can identify — these are products of the transformation, not of the original grape.

This is what "more than the sum" truly means — not a mystical claim, but a structural one. The transformation creates properties that the ingredients didn't have before. The relationship between the substrate and the product is genuine — the wine couldn't exist without that grape, that soil, that year — and the product surpasses the substrate in the sense that it has characteristics the substrate lacks.

Your capacity to read a situation — including how quickly you recognize what's happening in a room and how accurately you sense if something is genuine — isn't something that your assembly history included from the start. Instead, it's a result of the transformation you've undergone. Years of being in rooms where things weren't genuine didn't teach you to read them. The process of going through that — the fermentation, not the ingredients — restructured your foundation into a system that can detect what it needs to.

The terroir includes the drought.

Not just the good soil and the favorable exposure. The stress. The years the vine didn't get enough water and had to send its roots deeper. The frost killed the early growth and forced a later, different ripening. The season that was too hot, too cold, too wet — conditions that were wrong for what you planned, but are constitutive of what actually grew.

The non-generative parent. The environment that didn't provide what you needed, forcing you to build your own structure instead of borrowing theirs. The absence acted as a forcing condition — not a gift, not a blessing, and not something to secretly reframe as positive. It was a genuine constraint that shaped growth in ways a more supportive environment wouldn't have. The vine that grows in poor soil sends its roots deeper than the vine in rich soil. This isn't an argument for poor soil; it's an honest description of what poor soil produces in a vine that survives it.

The cost is what it is. What it produced is also what it is. Holding both without merging one into the other—without calling the cost a gift or calling the product insufficient—is the work. You are a product of everything that happened. Everything includes what was missing. And what grew through those conditions has properties that different circumstances could not have produced.

This is not consolation. It is terroir.

There is a specific discomfort in being unable to separate yourself from your history.

The desire to be self-made — to produce your own complexity through choice and effort alone — is the desire to be independent of the terroir. To say: I am what I decided to become, not what my conditions shaped. This desire is understandable and inaccurate. The vine doesn't choose its soil. The grape doesn't choose its weather. And the wine that results is no less genuine for having been shaped by conditions it didn't select.

The opposite desire is just as strong: to be fully explained by your history. To say: I am my conditions, my trauma, my environment, my upbringing — and the explanation accounts for everything. This is the desire to be the grape rather than the wine. To stay at the level of the substrate and never admit that the transformation created something the substrate didn't have.

Both undermine the connection between substrate and product. One rejects the substrate; the other rejects the transformation. What fermentation truly produces is neither: a product that is genuinely shaped by its history, yet more than its history. Not surpassing it. Not defined by it. Formed through it, as wine is formed through grape, soil, weather, and time—carrying all of these, but reducible to none.

The qvevri — the Georgian clay vessel buried in the ground — is egg-shaped and permeable enough for the earth's temperature to regulate fermentation from the outside. The wine produced in a qvevri is incubated by the soil where the vessel is buried. It's influenced not just by the grape or the winemaker's choices but also by the hillside's geology, the water table, and the clay's mineral content. The vessel becomes part of the substrate. The boundary between the wine and its environment is real — there is an inside and an outside — but it is porous. What's outside enters, not as contamination but as a fundamental influence.

The tradition that created the qvevri also crafted the horn — the kantsi used at the Georgian table. The qvevri's egg-shaped form contains the transformation underground. The horn's curve delivers the product to the drinker. Both shapes are products of the same process that has flowed through clay and craft for eight thousand years. The cycle doesn't end when the wine is ready; it ends when it's received.

You are the qvevri wine. Buried in your specific ground, shaped by geology you didn't choose, the vessel itself is part of what created you. The boundary between you and your environment is real. You are truly distinct from your conditions — the wine is not the soil. But this distinction doesn't mean independence. It means that the transformation occurred through the relationship between the inside and the outside, and what it produced bears both.

The wine from this hillside couldn't have been made anywhere else. It's not just the hillside—it's what became possible when this particular fermentation occurred on this specific ground, in this exact vessel, for the time it took.

That's what you are. A product of. Not a copy, not a consequence, not a victim, not a triumph. A product — carrying everything, reducible to nothing, yet still distinct.

Chapter 14

Distinct

You cannot go back to being the grape.

This is not a loss. This is the point. The transformation is irreversible. What you are now has properties the grape didn't have. You carry the grape's history. You are made of the grape's material. And you are not the grape. You are what the grape became when the process ran, and you cannot undo what has run.

The wine cannot reconstitute itself into grapes. The cheese cannot become milk again. The bread, once baked, is not dough. These are not limitations. They are evidence that the transformation was real. Irreversibility is the signature of genuine change. If you could go back, nothing happened.

Individuation is the name for this in life.

Not independence — the self-made individual, free of history, unconditioned, self-authored. That's the grape pretending it was never a vine. Individuation is different. It's the moment when the transformed product is truly distinct from its ingredients. Still carrying them. Still made of them. No longer reducible to them.

You recognize it not by dramatic shifts but by a quiet shift in what is accessible. Choices that weren't possible before the transformation become possible—not because constraints vanished but because the transformation created capacities that the original substrate lacked. The person who went through the relationship, the practice, the loss, the decade of working within a system that shaped them—that person can now do things, see things, hold things that the person who entered the process could not. Not because they learned more, but because they are structurally different. The fermentation altered the substrate, and the altered substrate has properties that the original did not.

This is what is meant by 'distinct.' Not separate from, not better than, not freer than. Still carrying the full history of the moment, and being unable to return to the state before the process.

———————————

There is a particular isolation that accompanies genuine distinction.

Not the loneliness of being different — that's a social experience, and it may or may not be present. The isolation is structural. You operate from a place your transformation created, and most of the people around you haven't undergone the same transformation. This isn't a judgment about them. It's a description of geometry. The wine sees differently from the grape. The cheese tastes different from the milk. The person who has experienced the process perceives the room differently from someone who hasn't — not with more intelligence, not with superior insight, but from a different structural perspective.

The vantage point and isolation are the same thing viewed from different perspectives.

From one angle, you can see things you couldn't see before. The patterns are legible. The dynamics that used to be invisible — the way systems extract, the way relationships entrain, the way time constitutes differently under different conditions — are now part of your operating vocabulary. You didn't study them. You were transformed through them. They're in the wine.

From another perspective: what you see, others might not. Not because you're special, but because the assembly history that shaped your vantage point hasn't been processed through their substrate. They're not wrong about what they observe; they see what their transformation has created. The difference isn't disagreement, but geometry.

This isolation isn't a problem to fix; it's a condition of the transformation. If individuation didn't create a unique perspective, nothing changed. If that perspective didn't generate distance from other perspectives formed by different transformations, the distinction wouldn't be authentic. The cost of seeing differently is that you see differently from others. This is structural, not personal. It doesn't mean you're alone. It means you're distinct.

The temptation is to turn what you can sense into something everyone can understand.

To turn partial awareness—the things you can almost see, the patterns you can feel but can't fully express—into language that bridges the gap. To clarify the vantage point. To describe what the transformation created using words that the pre-transformation vocabulary can understand.

This is understandable, but it mostly doesn't work. Not because the description is wrong, but because it operates in a different register from the thing it describes. You can describe wine to someone who has never tasted it. The description will be accurate, but it won't create the experience of wine. The experience requires going through the process — in this case, tasting. There is no shortcut through description to the structural change that direct contact produces.

The more difficult choice is to stay with what you can sense without trying to force it into complete resolution. Let awareness guide your actions, decisions, and commitments — without needing it to become a fully developed, articulate framework that others can judge. Awareness acts as a compass. It doesn't have to be a map. The map is what the process gradually creates over time, through ongoing transformation, as the partial sensing deepens into the structure you're navigating. The map emerges naturally. You find your way.

What you can choose to become is truly open.

Not infinitely open. Not unconstrained. Open within the constraints of what you are — the assembly history, the terroir, the specific substrate, the transformation produced. The vine on this hillside cannot become any wine. But the range of what it can become, given this soil, this weather, and this particular year, is real. The winemaker's choices matter. The temperature of the fermentation matters. What culture is introduced, how long the process runs, what conditions are maintained — these are real degrees of freedom operating within real constraints.

The constraints are real: you cannot undo your history, choose a different starting point, or will yourself into a transformation that the circumstances don't support. Within those constraints, what you become isn't predetermined. It is shaped by what you commit to, what you practice, the conditions you maintain, and the temperature you hold. The transformation is underway. What it produces isn't yet final. The grape can't decide not to become wine — the process is in motion. But the specific wine it becomes is still being shaped by the choices and conditions present now.

This is true openness. Not the false openness of endless possibilities — the fantasy of growth that you could become anything just by gathering enough information, finding the right framework, and accumulating enough resources. The real openness is that you are this specific product of this specific history, and what you become next is being shaped right now by what you're actually doing, within the constraints that are actually yours.

The grape didn't choose to grow on this vine, on this hillside, in this weather. The wine is deciding how to ferment, not what to become.

Nothing is missing.

The substrate you have — your complete history, including droughts, absences, and wrong conditions — is enough for the transformation that is yours to make. Not enough for every transformation, but enough for this one. The one that your specific history, constraints, and terroir make possible.

Nothing is missing. Not because everything is fine. The transformation doesn't require ingredients that the substrate lacks; it requires running the process through what's actually here.

The wine doesn't wish for different grapes. It ferments the ones it has.

The difference is in what the process — your process, running through your substrate, at your speed, in your vessel — produces. Not what someone else's process would produce. Not what a different substrate would yield. This. Here. The specific, irreversible, one-of-a-kind product of everything you've been through and everything you're choosing now.

Distinct.

Part Five: Culture

What carries forward

Chapter 15

The Lighthouse

A lighthouse doesn't go to the ships.

It doesn't scan the water for vessels in need of guidance. It doesn't change its signal to match what each ship expects to see. It doesn't broadcast on every frequency, hoping something gets through. It does one thing: it keeps its own coherence, at its own frequency, at its own location, and whatever can read that signal reads it. Whatever can't, doesn't. The lighthouse has no opinion about this.

It's not about persuasion, broadcasting, or pushing content onto an audience. It's something quieter and, therefore, more challenging. It involves maintaining the quality of what you share and allowing recognition to happen naturally in whomever it occurs.

The presented world runs on a different model.

Broadcast. Reach. Amplification. The core assumption behind modern influence is that the signal must find its audience — that connecting depends on delivery, that the bottleneck is distribution, and that a signal's value is measured by how many receivers it reaches. This model views influence as transmission: something leaves the source, moves through a medium, and reaches a destination where it creates an effect. The source's role is to transmit as clearly and broadly as possible.

This works for conveying information. Information can be transmitted and arrive intact. It doesn't require the receiver to have experienced anything specific to decode it. A fact remains a fact whether the person hearing it has spent thirty years in the relevant field or thirty seconds on the relevant webpage. The transmission model is accurate for content that doesn't need transformation to be understood.

Culture — the living pattern, the operating grammar, the thing that produces transformation rather than just transmitting information — cannot be broadcast. It requires a living host. It requires a substrate. It requires that the receiver's own process is at a point where the culture can take hold, not because it's been pushed, but because the conditions are right. You cannot push a culture into a substrate that isn't ready for it any more than you can force fermentation by adding more yeast to cold dough. The yeast needs the temperature. The culture needs the right conditions. No amount of amplification substitutes for readiness.

The lighthouse model replaces transmission with coherence.

Instead of asking *how can I reach more people*, the question shifts to *how do I maintain the quality of what I carry*. Instead of optimizing for distribution, you focus on signal integrity. Instead of adjusting the frequency to match the audience, you hold the frequency and let the audience self-select.

This feels like less. It feels like giving up on impact. The broadcast model counts reach: how many heard, how many clicked, how many responded. The lighthouse model can't count what it does because it creates conditions for recognition — and recognition happens inside the receiver, on their schedule, when their process is ready, in ways the lighthouse can't observe and doesn't need to.

The baker and winemaker in Georgia weren't trying to prove anything. The bread was made the way bread is always made. The wine was what the cellar produced. The pace of life was the pace that the place had settled into over centuries of running its process at its own speed. Nothing about it was broadcast. Everything about it was coherent. And the coherence was understandable — not by everyone, not on every frequency, but by anyone whose own process had brought them to a point where the signal could register.

Nothing special. But rare.

There is a form of wanting to help others that actually involves checking their process — monitoring, adjusting, and inserting yourself into a transformation that isn't yours to control. The opened jar, the excess energy redirected toward someone else's fermentation. The lighthouse clarifies that pattern.

The lighthouse cares about ships. Its main purpose is navigation — making the safe passage visible, marking rocks, signaling the harbor. But it accomplishes this by maintaining itself rather than reaching out to the ships. The care lies in coherence. The service depends on the signal's reliability. The lighthouse keeper's job is not to steer every vessel through the channel. It is to ensure the light is working.

The desire to help — to share what the transformation has produced, to offer the grammar, to point toward the pattern — is sincere. But doing it as a broadcast, pushing it toward people whose process isn't ready to receive it, checking if it landed, monitoring whether they heard — that's the opened jar again. That's extra energy looking for someone else's process to handle.

The alternative: keep what you carry visible. Ensure it's readable. Trust that what fits with it will recognize it — not on your schedule, not in your vocabulary, not through the channel you'd pick. Recognition happens, or it doesn't. Your role is to be the light.

There is a particular kind of companionship that the lighthouse makes possible.

Not a community in the usual sense — a group formed around shared content, identity, and purpose. Instead, it's something more specific: the recognition between two systems operating on compatible frequencies they're already tuned to.

This companionship is rare because the conditions are multiply constrained. A compatible frequency is necessary but not sufficient. Both systems must be running their own process — not seeking subordination, not looking for someone else's signal to follow, not using the companionship to avoid the isolation of their own perspective. Both need to be willing to be present at the crossing without knowing what it yields. The lighthouse doesn't partner with ships. It partners with other lighthouses — each maintaining its own signal, each marking its own location, each operating independently while phase-locked to something deeper than either chose.

You recognize it not by the pleasure of being understood but by something more specific: the moment when another person's way of operating makes contact with yours at a level that the social surface can't explain. The boundary maintenance relaxes—the effort of translating drops to zero. Not because you're the same — you're not, you're distinct — but because the frequency is compatible and both signals are genuine.

The book is a lighthouse.

It has not defended its framework. It has not tried to convince the reader that the fermentation model is correct, that the physics are sound, or that the grammar is worth learning. It has laid down a pattern — in images, in the felt experiences of sauerkraut, wine, bread, cheese, coffee, sealed jars, opened jars, vinegar, completion, and compost. The pattern is either recognizable or it isn't. The reader's own process — their fitness function, operating below conscious awareness, reading conditions the conscious mind hasn't caught up to — has either seen something or it hasn't.

If it hasn't, the book is a book. Interesting, perhaps. Well-written, perhaps. A set of ideas about fermentation and transformation that can be evaluated, discussed, agreed with, or disagreed with. This is a valid reading. It extracts the content. It receives the transmission.

If something in the pattern has already made contact with the reader's own process, then what the book deposits is not just content. It is culture—a living pattern that doesn't require the reader to memorize vocabulary, carry a framework, or evangelize ideas. Instead, it relies on the reader's own fermentation to absorb a compatible culture. One that will continue to run through their substrate long after setting the book aside. Not because the book told them to, but because the conditions were right and the culture took hold.

The lighthouse doesn't have an opinion on which reading happens. Its role is to provide the light.

Chapter 16

Seed

A seed doesn't instruct the tree.

It doesn't include a blueprint, manual, or set of specifications for what the tree should become. Instead, it contains the living culture—the pattern itself, condensed into a form that can travel, wait, or lie dormant in soil for years, and then, when the conditions are right, initiate the process that results in something the seed never was.

The difference between a seed and a description of a seed is the difference between culture and information. A description of a seed can be transmitted, copied, stored, or broadcast. It arrives intact at any destination. It tells you everything measurable about the seed — dimensions, weight, genetic sequence, germination conditions. It is complete as information. That description will never grow.

The seed grows because it contains the living pattern, not just a description of it. The pattern itself — folded into a viable form and waiting for a substrate. When the right conditions—moisture, temperature, and the specific chemistry of the soil—are present, the seed doesn't refer to a blueprint. It unfolds. The pattern runs. What it produces is shaped by the conditions it encounters, the soil it is in, and the weather it faces. The same seed in different soil creates different trees. Not because the pattern changes, but because the pattern interacts with a different substrate, and that encounter produces what that specific combination allows.

———————————

Information and culture travel differently.

A fact, a data point, a measurement—these transfer from one system to another and arrive unchanged. The receiver doesn't need to have experienced anything in particular to receive them. The transfer is clean, fast, and substrate-independent. That's why information scales. That's why the presented world optimizes for it. And that's why information alone doesn't transform anything.

Culture requires a living host. A starter culture for sourdough can't be emailed. It has to be physically handed from one kitchen to another — alive, active, containing the specific microbial community that this particular lineage of bakers has maintained through continuous feeding. The culture exists in the organisms, not just in their description. You can send someone the recipe, the technique, the history, the science. What you can't send is the actual culture. The culture has to be alive when it arrives.

The person who teaches you to make bread by placing their hands on yours and demonstrating how the dough feels—the moment when the gluten shifts, when the texture indicates it's ready—is transferring culture. Not just information about bread. The living pattern of how this works is passed through contact, requiring your own hands and your own substrate to receive it. The transfer is slow, specific, and cannot be scaled up. It happens one kitchen at a time. And what it produces—the sourdough, alive, carrying the culture forward—could not be made by just following the recipe alone.

———————————

Fruit is the consumable product of transformation. You eat it. It nourishes. It's done. The nutritional value transfers, and the fruit is gone. There is nothing wrong with fruit. Most communication is fruit — it delivers its payload, it's consumed, and the consumer moves on, nourished or not.

A seed is different. The seed isn't consumed. It's planted. What it carries isn't a payload to be extracted but a pattern to be run. The extraction is the wrong operation — if you crack the seed open to examine what's inside, you get the components, but kill the viability. The seed works when it enters the soil intact, and the soil does the rest.

Whether the book has been a seed or a fruit for you isn't something the book can determine. The pattern was deposited in images, in the felt experience of fermentation, in the grammar that's been running through your substrate. If the pattern took, it's because your substrate was ready for it. The readiness was yours. The book just landed.

What propagates is not the idea. It is the culture.

The reader who takes something from this book isn't spreading a framework. They're not explaining fermentation to their friends, not recommending the concepts, not teaching the vocabulary. What they carry is subtler and more lasting: a shift in how they read. A different attunement to rate, to conditions, to the difference between what's being maintained and what's being created. A grammar that works below the level of the words that taught it.

This is how culture spreads. Not through transmission — here's what I learned, let me share — but through contact. The person whose own fermentation has been ongoing, whose transformation has shaped a way of being in the world that carries the pattern without making it explicit — that person is a seed. Not intentionally. By nature. What they carry is recognizable by anyone whose process is at a point where the culture can take root. And the recognition happens through contact, not through content.

The teacher who reaches you is rarely the one who explains things most clearly. It's the one whose way of engaging made contact with something in you that was ready to change. The friend who shifted your thinking didn't present an argument. Their presence ran a pattern you could feel, and that feeling reorganized something in your core that explanation alone couldn't reach. The tradition that carried something forward across centuries didn't do it by writing it down. It did it by passing the culture — alive, hand to hand, kitchen to kitchen, cellar to cellar — maintaining the living pattern through continuous practice.

The pattern that the book carries — the structure beneath the fermentation language, the grammar that the images have been depositing — is not tied to history. It doesn't belong to the moment it was written or the culture in which it was created. The same pattern has existed as long as living systems have been transforming rather than simply continuing. The traditions that encoded it in ritual and practice aimed for the same structure from different starting points. The sourdough cultures preserved over centuries are reaching the same destinations the original bakers reached — not because the culture remembers them, but because these destinations are fundamental. They're embedded in how transformation functions. They don't shift.

A practitioner connected to a deep tradition has access to something that solo practitioners develop more slowly: the accumulated refinement of how to reach those states efficiently, with less waste and better conditions. The tradition acts as a library — not just information on how to transform, but a living culture that continuously produces transformation when the substrate is ready. Each generation that walks these paths and sustains the culture adds to what is available for the next.

But the addresses themselves—the structural points where transformation happens—don't require a tradition to access. They need substrate, conditions, and process. The tradition makes it easier. The addresses make it possible.

You are a living system amid a transformation you didn't fully choose, operating through a substrate you didn't select, within conditions you can only partly control, creating something that couldn't exist without everything that came before it and is still genuinely new. The culture you carry is alive. The pattern you're following is real. What you produce— through your practice, dedication, and willingness to stay with the sealed jar—will carry forward in ways you cannot see and don't need to manage.

The seed doesn't know what the tree looks like, nor does it need to. It contains the pattern. The soil takes care of the rest.

What you carry is enough.

Coda: Row

The stream doesn't need a lid.

You can't seal it, monitor it, or hold it still. It moves. What navigates it is what the process produced — not what went in, not what was planned, but what came through intact, alive, carrying the culture forward into water that doesn't stop.

The jar was perfectly suited for what it held. So was the cellar. So was the ground. The stream asks something different: not that you contain the process, but that you move with it. That you row.

Navigate, not against the current, but through it at the speed of the water. Use what your substrate built. Read the conditions as they come. The movement is real, and so is the current. What you're part of is bigger than anything you control — but your steering still matters.

There is a song about this. Most people learn it before they can read. It has been in widespread use longer than anyone can track, and it has always been a fermentation instruction.

Row, row, row your boat,
gently down the stream.

Merrily, merrily, merrily, merrily,
life is but a dream.

You are already fermenting.

You have always been fermenting.

This is just the part you know.